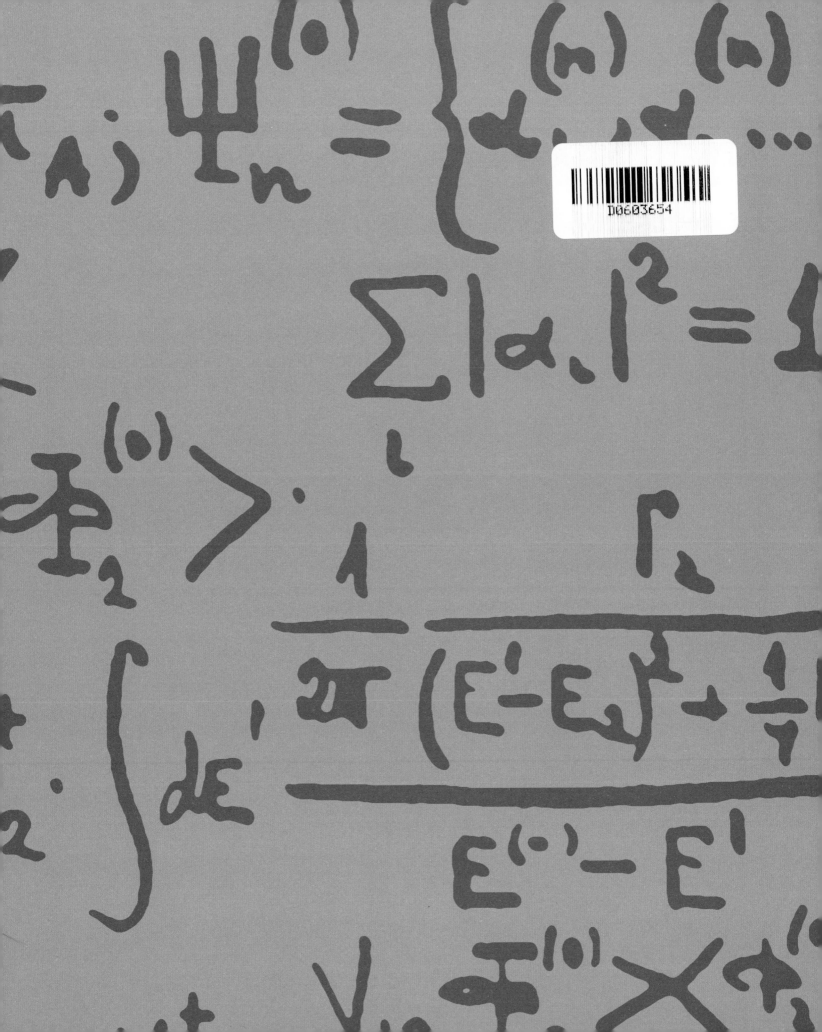
D0603654

MATHEMATICS

AN ILLUSTRATED HISTORY OF NUMBERS

PONDERABLES

100
BREAKTHROUGHS
THAT CHANGED HISTORY
WHO DID WHAT WHEN

MATHEMATICS

AN ILLUSTRATED HISTORY OF NUMBERS

CONTRIBUTORS

Richard Beatty • James Bow • Mike Goldsmith • Dan Green
Tom Jackson • Robert Sneddon • Susan Watt

Edited by Tom Jackson

SHELTER HARBOR PRESS

NEW YORK

Contents

Introduction

IS IT A SCIENCE? IS IT AN ART? WELL PERHAPS BOTH, PERHAPS NEITHER. MATHEMATICS IS A SUBJECT SET APART FROM ALL OTHER HUMAN ACHIEVEMENTS, THE INTERFACE BETWEEN INTELLECT AND IMAGINATION, WHERE THE REAL AND UNREAL ARE PRECISELY CONFIGURED.

An Arabian proof of the Pythagoras theorem uses a graphical approach to show the relationship between the lengths of the sides of a triangle.

Math started as a means to record wealth and divide up land. Most ancient math records, such as this 4,000-year-old tablet, are lists of transactions.

The thoughts and deeds of great thinkers always make great stories, and here we have one hundred all together. Each story relates a ponderable, a weighty problem that became a discovery and changed the way we understood the world and our place in it.

History is generally a rousing story of change, as ideas rise and fall, cultures dominate and then recede, one ponderable is replaced by another. Not so in mathematics. Once a mathematician has proven something to be true it cannot be disproven. Contrast the Earth-centered universe championed by the classical astronomer Ptolemy (and accepted as truth for the best part of 1,500 years) with the geometric techniques he developed to help orchestrate its motions. Ptolemy's universe is now a byword for misguided thinking. However, the Alexandrian's geometry worked then and it works now, forming the basis of trigonometry (so now you know who to blame).

THE FIRST PONDERABLE

The history of math is not one of brave new ideas conquering old, striking them from the record. Rather it is the story of how old but venerable truths are joined by newer ponderables, accreting gradually to the body mathematic.

Computer code is math writ large. When binary numbers of 1s and 0s are too long even for computers to remember, they get converted into hexadecimal numbers, which count from 1 to F!

Math has hidden depths. This table shows how turning the geometry of lines into algebra created a way of describing natural changes—from the growth of plants to the fall of stockmarkets.

The story begins with the number one, yet infinity is not its end, far from it. The term *mathematics* is derived from the Greek word for "knowledge." Practically every thing we know—rather than believe—begins with its quantification, its expression as numbers. The first ponderable is did those quantities already exist or did we have to invent them for our own purposes? If math is a construct of the human mind, it must be innate since the same number systems appear time and again in isolated cultures. The Maya invented the concept of zero independently of Babylon and Hindu India, and as far as we know these cultures never traded ideas—or anything else. Similarly the binary symbology of the I Ching of China is reflected in the math of Ifá divination hailing from the Niger River Valley.

Could it be that math reflects the patterns of reality, some manifest, others hidden? When not being cited for his work on right-angle triangles, Pythagoras is remembered as the first person to anchor math to a natural phenomenon, when he revealed the relationship between the length of a string and the musical tone it produced when plucked. As you will see, the orbits of heavenly bodies, the accrual of wealth, the inner mechanisms of matter, computers, and political strategies, and even the roots of beauty itself follow paths, lined with numbers. Let us begin that very journey.

FIELDS OF MATH

Math can be applied to anything—with varying degrees of success—and so defining it in terms of its applications results in a lot of confusion. It is like explaining the telephone by reading out the phone book. Even a theoretical basis for dividing up math requires a lot of compromise due to the huge potential for cross-over between subjects. In the simplest terms math tackles quantities, essentially different ways of counting; number structures, the patterns and linkages within; space, the characteristics of shapes and surfaces; and finally understanding change by tracing dynamic systems from instant to instant. Some specific fields are:

ALGEBRA
Investigating the relationships between numbers by replacing real, yet variable, numbers with generalized terms, most frequently x and y.

GROUP THEORY
The characteristics of a number group formed when members of a set are converted into a series of results by the same operation. Like numbers, such groups are often composites of simpler ones.

ORDER THEORY
The study of overarching principles that can be derived from the way any numbers, quantities, or other mathematical elements can be seen as "being less than," "more than," or "before" and "after".

NUMBER THEORY

ARITHMETIC
Manipulating quantities by operations such as adding, subtracting, multiplication etc.

MATHE

APPLIED MATHEMATICS

PROBABILITY
The mathematics of chance, which calculates the likelihood of one event given the initial conditions and possible outcomes.

STATISTICS
Revealing significant insights from mathematical samples taken from the real world.

FLUID DYNAMICS
Modeling the motion of liquids and gases based on quantifiable characteristics of the material—viscosity, velocity, pressure, etc.

CRYPTOGRAPHY
The math of codes from the binary numbers used to control computers to the encrypted messages sent between them.

GAME THEORY
Using math to evaluate real-world strategies that minimize losses and maximize gains based on probability as well as the behavior of opponents and cooperating partners.

TOPOLOGY

A form of geometry in which only the connections within an object are important; the lengths and angles are free to change without altering the fundamental characteristics.

TRIGONOMETRY

The relationship between the lengths of the sides of a triangle and its internal angles on planes as well as on convex and concave surfaces.

FRACTALS

Applying geometry and topology to the real world reveals "rough" surfaces and self-similar objects that are described in terms of fractions of dimensions.

DIFFERENTIAL GEOMETRY

Using geometrical techniques to explore dynamic systems or to apply calculus to complex geometry.

ALGEBRAIC GEOMETRY

Using algebraic expressions to describe the edges and surfaces of geometric objects.

GEOMETRY

MATICS

PHILOSOPHY OF MATHEMATICS

LOGIC

Tackling why 1+1 = 2 and other questions that form the foundations of math.

SET THEORY

The study of sets of numbers, how they relate, compare, or overlap and whether they are infinite or not.

GRAPH THEORY

The study of the connections between lines, shapes, and other objects.

FUNCTIONAL ANALYSIS

DIFFERENTIAL EQUATIONS

Dealing with mathematical expressions that involve a function and its derivatives. A function is a defined operation performed on an input variable; the derivative is a snapshot of the function's effect.

INFORMATION THEORY

Measuring the units of information or data, like that stored in computers and transmitted between them.

CALCULUS

A field of math that describes a continuous change—in a real-world motion, or the swoop of a purely notional algebraic curve—in terms of infinitely small and instantaneous steps.

CHAOS THEORY

Describes how dynamic systems can have widely differing outcomes from only slightly varying initial conditions; why one tiny change can make a big difference.

PREHISTORY TO THE MIDDLE AGES

1 Learning to Count

Math starts with counting; even the most brain-bending trip into the abstract landscape of numbers is meaningless without 1,2,3... There is nevertheless one seemingly simple problem that math has yet to solve: did humans invent numbers or were they there already?

Leopold Kronecker, a 19th-century German mathematician, was only half joking when he said, "God made the integers; the rest is the work of man." The integers he referred to are the natural, whole numbers 0 to 9; 10 onwards are just using them again. (The word *integer* is from the Latin for "untouched.")

So are numbers as much a part of nature as atoms and forces? The advent of counting is lost in prehistory. While it is likely that pre-human ancestors were able to recognize small quantities, twos and threes etc., it has been suggested that formalized counting would have only arisen when there were larger numbers of things to count. Stone Age people carried several items in their toolkit; knowing how many exactly would have been useful. Precise values were recorded as simple tally marks, and those on rock and bone have survived. (The Moksha, an ethnic minority in western Russia, have traditional numerals that are barely amended tally marks, which are thought to have been in use since prehistory.)

An explosion in counting—and records of its results—occurred when humans gave up a nomadic, hunter-gatherer life style and began to live as settled farmers. At this point people began to accrue livestock and the paraphernalia of civilization—items of value that were worth counting. And once counted, these quantities were compared with others and then combined, traded, and multiplied—mathematics was born.

The Ishango bone, carved from a baboon's fibula in Central Africa 20,000 years ago, is one of the oldest mathematical tools on record. Instead of being a simple account of quantities, the tally marks on the bone appear to have been used for calculations working in a counting system based on the number 12.

COUNTING AT A GLANCE

We often use numbers in an inexact way, saying *a few*, *several*, or even *billions*. In these cases, the precise figure is not important and we do not have the time—or means—to count the objects in question. However, when it comes to being exact, our brains appear to have an innate maximum. Look at these stones. There are six, of course but your brain probably recognized them as two sets of three. Cover one and look again. The human brain appears to be able to recognize a maximum set of five. Counting higher than this requires joining smaller sets together.

2 Positional Notation

WITH FINGERS AT OUR DISPOSAL COUNTING UP TO TEN IS EASY ENOUGH. HOWEVER, COUNTING ANY HIGHER requires a different strategy. Today, we use a positional counting system that counts in 10s. The earliest positional notation hails from Babylonia, 5,000 years ago—and it counted with 60s.

The Babylonians used a wedge-shaped reed to write in wet clay, which dried out into solid tablets. It is little wonder, therefore, that their numerals were composites of wedges. They had inherited a sexagesimal (base 60) system from earlier civilizations. There is a lot of sense to it. Sixty can be divided by 1, 2, 3, 4, and 5—compare that to 10. The sixty minutes in an hour and 360 (6x6) degrees in a circle are handovers from Babylonian math. Like the 0 to 9 we use today, the value of a Babylonian numeral could represent units, tens, or 60s depending on where it appeared in a string of numerals—reading Babylonian is best left to the experts. Compare this to the later Roman system, which had fixed values for numerals irrespective of position: LXI is 50+10+1 or 61.

1	𒀸
2	𒐖
3	𒐗
4	𒐘
5	𒐙
6	𒐚
7	𒐛
8	𒐜
9	𒐝
10	𒌋

The first ten digits of Babylonian numbers. There were also symbols for 40 and 50.

3 The Abacus

MANY BELIEVE THAT THE ABACUS PREDATES WRITTEN NUMBERS. It has been suggested that Babylonian numbering was invented as a means to record the numbers calculated by moving beads.

Although it is all laser-guided code scanners and self-service checkouts these days, it was not too many decades ago that a shopkeeper might have used an abacus to tally up groceries. In many parts of Asia merchants still rely on the frame of rods and beads to perform complex calculations at impressive speeds. The counting frame abacus is perhaps not quite as ancient as it might be imagined, being a fusion of the counting tables of the Near East with the counting rods of the Far East. The word *abacus* derives from Arabic for "dust" which probably alludes to a frame of sand with columns of pebbles or some other kind of counter arranged within. In China, the counters were stacked on rods, a bit like a toddler's puzzle. (Chinese numerals from around the 3rd century BC even resembled upright rods with stacks of discs.) It was not until the 16th century that the two approaches were combined into a handy frame.

A Chinese abacus has two decks for counting in 10s (using one upper bead for 5) or in 16s (two uppers make 10 adding up to 15 with the lower deck). Hexadecimal counting is used for calculating traditional Chinese weights which are subdivided into 16 units.

4 Pythagoras Theorem

ONE NAME FREQUENTLY TOPS SURVEYS OF THE WORLD'S MOST FAMOUS MATHEMATICIAN: PYTHAGORAS. That is quite an achievement for a man who may not have existed, was a prime suspect in a murder, and did not even formulate the theorem for which he is famous.

Beyond multiplication tables and the basic operations of arithmetic, the Pythagoras theorem is the most widely taught subject in mathematics classes. There is a certain clarity to it, which makes it simple to remember: $a^2 + b^2 = h^2$. For those readers out of reach of their school books and in need of a reminder, this equation tells us that if you square the length of the two short sides, or legs, of a right-angle triangle, they will add up to the square of the hypotenuse—the triangle's longest side. So if you know two lengths, you can always calculate the third.

Pythagoras was a hallowed figure in classical Greece, although much of his biography—if not all of it—is myth perpetuated by his devotees, Plato among them.

Practice before proof

We name this theorem after Pythagoras of Samos, who lived in southern Italy about 2,500 years ago. The theorem had been known about for at least several centuries prior to this, but as far we know Pythagoras was the first mathematician to prove that it was true. Pythagoras traveled extensively as a young man visiting Egypt and Babylon and he may have reached as far as India. In these places he would have seen "his" theorem in action as a practical tool in surveying and construction. Egyptian surveyors habitually used ropes knotted at lengths of 3, 4, and 5 units. When all three were used to construct a triangle they always created a perfect right angle. The numbers 3, 4 and 5 form the first of an infinite set of "Pythagorean triples," three whole numbers that obey the theorem.

In 2002, the Pythagoras theorem was applied in a thoroughly modern context in a New York courtroom. Judges were instructed to give any drug dealer found operating within 1,000

A tomb painting from 1,400 BC shows ancient Egyptian surveyors measuring out wheat fields before the harvest. Each field would have perfect right-angle corners thanks to a rope system of Pythagorean Triples.

paces of a school gate a heavier sentence. But was that distance measured in so-called "Manhattan distance"—the straight-line journey through a grid-iron of city blocks—or a direct diagonal as calculated by the Pythagoras theorem? The court opted for the simplicity of Pythagoras.

Irrational philosophy

Pythagoras saw numbers as sacred: everything in nature could be described in terms of whole numbers. He had built up quite a following and led a secret society of mathematicians devoted to revealing the absolute truth of numbers. However, while his theorem sealed Pythagoras's legacy as a mathematician, it destroyed his philosophy. Hippasus, a member of the Pythagorean cult, pointed out that if a triangle had legs of 1 unit, then the length of the hypotenuse was $\sqrt{2}$. This square root was made up of an infinite string of numbers and therefore had no exact value (it is an *irrational* number). Faced with this devastating blow to his authority, legend has it that Pythagoras invited the problematic Hippasus for a fishing trip, and returned to shore alone...

THE PYTHAGOREANS

Although he was born on the island of Samos, located off the western coast of what is now Turkey, Pythagoras spent most of his life in Croton, a Greek colony in southern Italy. His followers, the Pythagoreans, were something of a cult. Only a select few—presumably those with the necessary mathematical rigor—were initiated through secretive rituals. The sect lived by the tenets of the Golden Verses, such as, "Search for the just measure, which is the measure that does not cause pain" and "Be kind with your words and useful with your work." The Pythagoreans also gave each number certain significance: 1 symbolized reason, 2 was the undefined feminine spirit, 3 was the sum of 1 and 2 and therefore masculine; 5 (2+3) was the most powerful of all. However, the Pythagoreans are also remembered for more outlandish practices. They were reputed to be scared of white cockerels and never touched beans. When the cult eventually lost the support of the people of Croton, Pythagoras' enemies came to kill him. It is reputed that they only caught up with him because the great master refused to flee across a bean field. He chose death over causing harm to the crops.

This simple diagram ably demonstrates the relationship between the squares of the sides of a triangle. It is also shown in an antique Arabian textbook. In the 19th century, finding more ingenious ways to transform the areas of A and B into C became a popular parlor game in mathematics circles.

5 The Rhind Papyrus

Ahmes claimed that the 2-meter papyrus gave "Accurate reckoning for inquiring into things, and the knowledge of all things, mysteries...all secrets."

THE RHIND PAPYRUS, NAMED AFTER HENRY RHIND WHO BOUGHT IT WHILE VISITING LUXOR IN 1858, gives us a glimpse into the mathematical world of the ancient Egyptians. The document is also sometimes referred to as the Ahmes Papyrus, after the scribe who copied it between 1650 and 1500 BC from a text that was written around two centuries earlier.

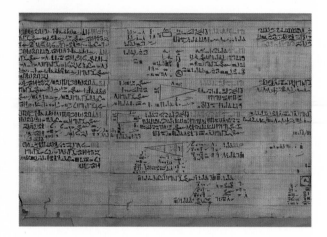

The ancient document presents reference tables to aid in calculations and solving more than 80 mathematical problems of the sort that would be encountered by the administrators of Egypt around 1650 BC, such as finding the volume of a granary. The papyrus also provides us with evidence that the Egyptians calculated the area of a circle using a formula that gave an approximate value for pi of 3.1605. Like other civilizations of the time, the Egyptians would have found this value for pi by comparing direct measurements of diameters and circumferences.

6 Zero

The use of a circle, or dot, to symbolize zero is seen in Greek, Indian, and, Chinese texts.

THESE DAYS WE MIGHT TAKE NOTHING FOR GRANTED, BUT ZERO HASN'T ALWAYS BEEN AROUND. The Babylonians were using it as a placeholder more than 3,000 years ago. Eventually, the Indians made it a number in its own right.

What we now understand as a zero first appears in Babylonian numbers as a pair of angled wedges. This zero represents an absence of a value within a larger number. (In 404, the zero stands for the absence of any tens in the number.) This type of zero turned up independently in Mayan numbers, but much closer to Babylon, Greek math based on geometry did not require a zero until the 2nd century BC, when Hipparchus introduced it for use in astronomy. A millennium later, Indian mathematicians made zero into a number like any other by showing that some expressions produced zero as the answer. This breakthrough paved the way for negative numbers, i.e. numbers that are less than zero.

7 The Math of Music

ANCIENT GREEK MATHEMATICIANS SAW NUMBER PATTERNS EVERYWHERE,
and not without good reason. Music, that special, harmonious subset
of sound, was found to be firmly rooted in math.

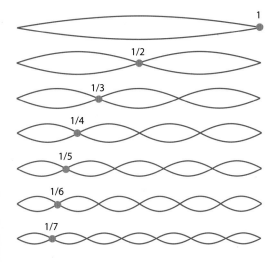

The wave of an oscillating string moves the air to create an analogous sound wave that we can hear. Dividing the wave by whole numbers creates a series of harmonics, the sounds of which combine to create a musical chord.

A story is told of how the Greek mathematician Pythagoras stopped to listen to the sounds of hammering from a blacksmith, and discovered that a hammer weighing half as much as another produced a note an octave higher as it struck the metal.

This event may never actually have taken place, but Pythagoras certainly conducted experiments into the relationship between the size of an object and the tone it produced, including plucking strings of different lengths and striking vessels filled with varying amounts of liquid to see how the notes changed. In doing so he established a mathematical relationship between object and sound.

Harmonics

Take two taut strings of the same material, one twice the length of the other, and pluck them. The short string vibrates with twice the frequency of the longer string and the resulting notes are an octave (eight tones) apart. The ratio of the octave, as Pythagoras discovered, is 2:1. If the string is a third of the original length, a ratio of 3:2 results. The interval, or difference in the notes produced, is a fifth. A ratio of 4:3 (from a quarter length string) produces an interval of a fourth. Sounding an octave, a fifth, and a fourth together produced a harmonious sound that was pleasing to the ear—a musical chord.

For the first time a natural phenomenon, sound, could be explained in terms of numbers, something that had never been done before. Pythagoras believed that musical harmonies must be reflected in the Universe as a whole, and that numbers and their relationships could explain all things.

MUSIC OF THE SPHERES

The ancient Greeks believed that the Moon, the Sun, the planets, and the stars were set in crystal spheres that revolved around the Earth and that these spheres made music as they turned. According to the Pythagoreans the distances between the planets had the same ratios as those that produced harmonious sounds in plucked strings. The closer spheres produced lower tones while those farther away moved faster and produced tones of a higher pitch. These tones blended together to form the music of the spheres that filled the heavens.

Ioannis Keppleri
HARMONICES
MVNDI
LIBRI V. QVORVM

In his 1619 book Harmony of the World *Johannes Kepler presented the relative pitches of the planets.*

8 The Golden Ratio

The proportions of the successive chambers in a nautilus shell approximate to the golden ratio.

IT IS OFTEN SAID THAT THERE IS BEAUTY IN MATHEMATICS, AND BY THE MIDDLE OF THE 5TH CENTURY BC, AND PROBABLY WELL BEFORE, it was known that there was also a lot of mathematics in beauty.

Despite the golden ratio being named after him, Phidias did not really stick to that proportion with his design of the Parthenon. It is slightly too tall to be golden, either due to surveying errors or because Phidias just preferred it like that.

The golden ratio is a number with perhaps more names than any other. It is called the golden section, the divine proportion, or simply *phi*. It relates to a mathematical means of dividing something into two unequal parts, generally with rather pleasing results. The golden ratio is often seen in the monuments and art of all ages. It is named phi after Phidias, the Greek architect who is said to have employed it in the proportions of his most famous work, the Parthenon of Athens, designed in the 440s BC. Euclid makes the first surviving record of the golden ratio in *Elements* in around 300 BC. However, its real wonder is not its presence—real and apparent—in human-made objects from credit cards to Leonardo da Vinci's Vitruvian Man, but its appearances in natural phenomena, from the growth of flowers and shells to the patterns within numbers themselves.

Finding the ratio

However you express it, what we call the golden ratio has a certain simplicity, correctness, and charm to it. Euclid described its application as "to cut in extreme and mean." A more mathematical depiction is to say if the golden ratio is x, then $x^2 - x - 1 = 0$. Putting it another way $x/1 = 1/x-1$. In words, the golden ratio is a proportion in which the "whole line is to the greater segment, as the greater is to the lesser."

Golden rectangles can be divided up into an infinite number of ever smaller golden rectangles by removing the square of the shortest length. In the terminology of Greek geometry, this makes the golden rectangle a gnomon, an object that preserves its shape as it grows (or shrinks).

A good example of the golden ratio is the credit card, a standard size the world over. In accordance with the golden ratio, the proportion of the short and long side is the same as the that between the long side and the sum of the short and long side. This makes the credit card a golden rectangle, a shape chosen for its balanced appeal—it is not too long or too wide. One way of checking if a rectangle is golden is to place two side by side, one vertical on its short edge, the other horizontal on its longer side. If the diagonal connecting the corners of the horizontal rectangle continues into the vertical one and meets its top corner, then you have two golden rectangles. Golden rectangles appear most often in architecture—such as the UN building in New York City.

Math meets art and nature

There is of course something prosaic about the golden ratio, at least to the unmathematically minded: its number. The solution to the algebraic expression $x^2 - x - 1 = 0$ is that x equals 1.6180339887... it continues without end.

Nevertheless, the ratio has a strong link with Western art, which can be traced in large part to the work of Luca Pacioli at the turn of the 16th century. Pacioli was a contemporary of Leonardo da Vinci and several of the maestro's drawings—including the most familiar version of Vitruvian Man—appear in Pacioli's 1509 work *De Divina Proportione*. This book establishes the geometrical basis for beauty using the number phi as its inspiration. For example, in the ideal proportions of the human body, the golden ratio was used to relate the height to the navel to the total height. Sadly actual measurements show that very few of our bodies are "ideal."

In the 20th century the golden ratio was sought in natural forms. For those who searched hard enough it was found in the proportions of leaves, the arrangement of buds and stems, even the trajectory of hawks going in for the kill. To some this showed a plan behind the structure of nature. To others it revealed that perhaps what we perceive as beautiful, or at least pleasingly proportional, is underwritten by the math of growth, which controls how structures can become larger without losing overall shape.

GOLDEN SPIRAL

A spiral that grows according to the golden ratio can be approximated from a set of golden rectangles. This is a variety of logarithmic spiral where the curving line diverges at a fixed rate. This kind of spiral is attributed to Jacob Bernoulli, its greatest proponent. Bernoulli asked for one to be engraved on his tomb, but the untutored stone mason gave him a non-diverging, more circular, Archimedean spiral instead!

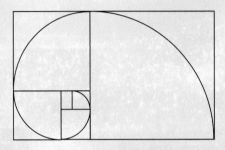

A golden spiral is hard to draw accurately. The best approximation is to draw circular arcs between the corners of the square segments inside each subdivided golden rectangle. (There are five in this example.)

9 Platonic Solids

ANCIENT GREEK MATHEMATICIANS SAW NUMBERS AS SACRED, EACH INFUSED WITH A SPIRITUAL QUALITY. Five was especially significant, a symbol of fertility and humanity's passion combined with reason. Backing up this belief was a fact: five is the maximum number of possible regular solids.

It may seem the antithesis of modern math but in the 4th century BC, numbers had personalities, and even the simplest calculations were steeped in meaning. This legacy of the great Pythagorean school of the 5th century was passed to the next generation of mathematicians, led by Plato, the towering Athenian philosopher.

In his 360 BC work, *Timaeus*, Plato describes five regular solids, or polyhedra. Regular means that all the edges were of equal length, all connecting at the same angle, and thus with faces of identical, also regular, polygons. The shapes are the tetrahedron, comprising four equilateral triangles; the cube, made of six squares; the octahedron, with eight triangles; the dodecahedron, which has 12 pentagonal faces; and the icosahedron with 20 triangular sides. Plato credits his Pythagorean forebears with the discovery of the solids, although his contemporary Theaetetus was probably the true discoverer of the octahedron and icosahedron.

Plato states what was already widely understood, that these were the only possible regular polyhedra. Plato then added a physical role to the shapes' mystical character making them the elemental building blocks of nature. As a result they have been known as Platonic solids ever since.

The Platonic solids as they appear in the Philosophia Pyrotechnia *by William Davisson in 1635, an early book about natural substances. Nowadays, the Platonic solids—or cosmic bodies—are little more than mathematical oddities.*

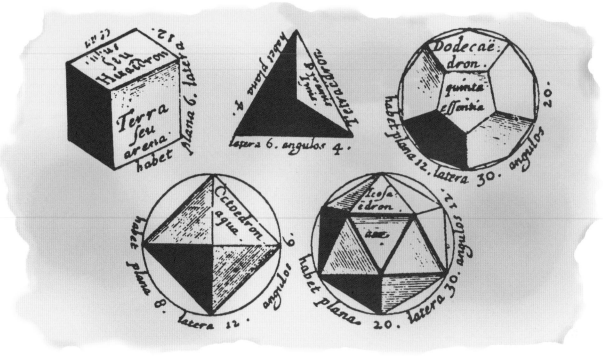

10 Logic

MATHEMATICS IS NOT POSSIBLE WITHOUT A BASIC ASSUMPTION OF HOW WE DERIVE ANSWERS FROM QUESTIONS. This is logic, an early formalized version of which was described by Aristotle in the 4th century BC. Much of today's math still relies on this way of thinking.

A 1570 Latin translation of Aristotle's work on logic. Aristotle set out 8 valid forms of syllogisms where a conclusion could be formed from two premises. In the 19th century, George Boole showed that two of these were in fact fallacies.

When we construct a mathematical equation we assume that its components will interact with each other in clear, predetermined ways. Thus the result we obtain will be repeatable—it will always be the same however many times we perform the calculation, as will other problems like it that rely on the same set of presumably correct assumptions. Therefore, at the center of mathematics: the assumption that if a specific cause or relationship produces a certain result it will do so in every identical instance.

Logic was defined almost 2,400 years ago by Aristotle, who set down the parameters of this philosophy of viewing the world around him in a collection of works known as the *Organon*, meaning "The Instrument." Central to Aristotle's logic was deduction. In his words, "A deduction is speech (logos) in which certain things having been supposed, something different from those supposed results of necessity because of their being so." What is *supposed* are the premises of the argument, and that which *results of necessity* is the conclusion. This process is known as a *syllogism*, from the Greek word for "inference." Syllogisms that result in incorrect conclusions are called fallacies. Aristitle showed that there are 256 possible versions, most of which are fallacies.

Aristotle's works were lost to Europe in the Dark Ages, but the original Greek texts were preserved in the Byzantine Empire. First translated into Arabic by Muslim scholars around 750 AD, the concept was revived in Europe in the 11th century. By the 19th century advances in mathematical logic began to pick apart syllogisms, showing their limitations. Nevertheless they remain a central part of set theory, which is relevant to everything from statistics to the concept of infinity.

ANATOMY OF A SYLLOGISM

The basic structure of a syllogism has three parts: the major premise, the minor premise, and the conclusion. For example:
• All mammals have hair (major)
• All people are mammals (minor)
• All people have hair (conclusion)
Both the major and the minor premises have something in common with the conclusion. The major term (mammals) is the predicate, whereas the minor term (people) is the subject. The hair is the so-called middle term. Syllogisms also use quantifiers. In the above example this is "all" but could be "no" or "some". Different combinations of quantifiers in the premises result in universal conclusions, which always apply, or "particular" conclusions which are valid only in certain cases. Like all deductive reasoning, syllogism can lead to fallacies because the veracity of the conclusion relies on the accuracy of the premises. Just one mistake leads to a string of false premises.

11 Geometry

THE ANCIENT GREEKS MAY NOT HAVE INVENTED THE FIELD OF GEOMETRY—Chinese scholars were working on its elements independently at around the same time—but they certainly established many of its basic assumptions and proofs.

The word *geometry* comes from the Greek words for Earth (*geo*) and measurement (*metry*). As the name implies, the Greeks were interested in measuring the elemental forms of nature. Practical applications of geometry relate to surveying techniques, mathematically determining length, area, and volume, but Greek scholars quickly realized that there were patterns and rules governing the shapes.

Around 300 BC, a Greek mathematician named Euclid of Alexandria gathered and expanded upon the principles of geometry in a thirteen-book treatise called *Elements*. Here, he set down a collection of definitions, axioms, theorems, and mathematical

Euclid's Elements *is the most translated, copied, and published secular book in history. These pages date from the Renaissance.*

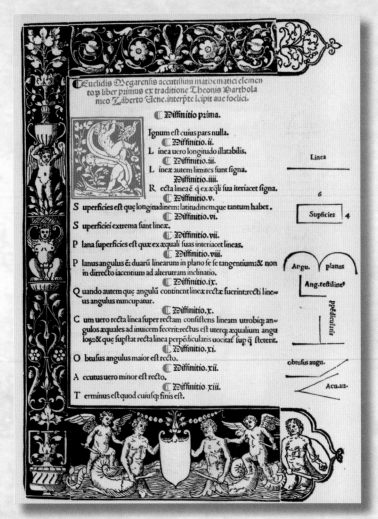

proofs that became the first principles of geometry and all elements of mathematics that derive from it. Such was the depth of his contribution to mathematics that Euclid is very much the "Father of Geometry."

Postulates and notions

Many of the theorems set out in Euclid were not originated by him. His contribution was to make them all conform to the same set of assumptions, or axioms. Among these were Euclid's five common notions: 1) Things that are equal to the same thing are also equal to one another. 2) If equals are added to equals, then the wholes are equal. 3) If equals are subtracted from equals, then the remainders are equal. 4) Things that coincide with one another equal one another. 5) The whole is greater than the part.

The five postulates are a little more geometrical: 1) A straight line segment can be drawn joining any two points. 2) Any straight line segment can be extended indefinitely. 3) Given any straight line segment, a circle can be drawn having the segment as radius and one endpoint as the center. 4) All right angles are congruent (i.e. can be transformed into each other). 5) If two lines are drawn which intersect a third in such a way that the sum of the inner angles on one side is less than two right angles, then the two lines inevitably must intersect each other on that side if extended far enough. (This final postulate is what is known as the parallel postulate and was later shown to be unprovable. It led to new forms of geometry based on a different set of axioms.)

The man and the work

Elements is the most influential textbook ever written, still in print after 23 centuries. The work survived due to Theon of Alexandria, who produced an edition in the 4th century AD. It proved to be an inspiration to the likes of Copernicus, Galileo, and Newton to name but a few world-changing thinkers. Nothing is known of Euclid himself. In fact it is only a few brief mentions by contemporaries and a statement by Proclus in his book *Commentary on the Elements* that lead us to assume that a man named Euclid indeed wrote the book.

COMPUTER GRAPHICS

At its simplest, computer-generated imagery (CGI) converts the complex shapes of nature (like a face) into a series of simple shapes. The overall shape is then defined by the smaller units, and can be modified with adjustments to their geometry. The idea stems from the work of mathematicians such as French-American Benoit Mandelbrot who, in 1974, showed that natural shapes conformed to fractal dimensions, and could not be measured but only approximated with traditional Euclidean geometry.

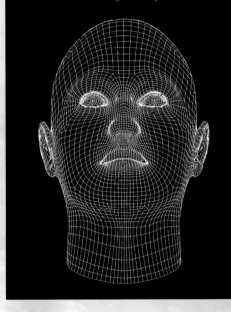

12 Magic Squares

IT IS PERHAPS NOT SURPRISING THAT SO MANY BREAKTHROUGHS IN MATH APPEAR THROUGH PUZZLES. One of the most ancient, the magic square, is first found in a seminal Chinese work—and is attributed to a highly numerate river turtle!

4	9	2
3	5	7
8	1	6

According to *Nine Chapters on the Mathematical Art*, a Chinese book from the 3rd century BC, the first magic square was given to humans by a river turtle. The so-called Lo Shu (river scroll) square has the digits 1 to 9 numbers arranged in a 3 by 3 grid. As in all magic squares, no number is repeated and in all directions, down across and diagonally through the center, the numbers add up to the same number, the magic constant. The Lo Shu is an order 3 magic square—and uses 3^2 digits. The order 1 square is rather uninteresting, while order 2 is not possible. Above that, there is an infinite number with the order n square using the digits from 1 to n^2. The magic constant can be calculated as equal to $n(n^2 + 1)/2$.

An order 4 magic square appears as a detail in Albrecht Dürer's 1514 engraving Melencolia I. *The artist is presenting his mathematical credentials.*

13 Prime Numbers

TO A MATHEMATICIAN PRIME NUMBERS ARE GLEAMING GEMS AMONG THE ENDLESS SANDS OF INTEGERS. A prime is a number that cannot be divided by another, other than 1 and itself. Other whole numbers are mere composites of the primes—made from two or more primes multiplied together.

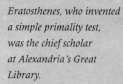

There is an infinite number of prime numbers, and as yet—despite millions of man hours spent in vain—no one has found a means to predict them. If there is a pattern to the primes, we cannot see it. We just have to search for every one of them. The concept of a prime is simple enough, but on the face of it, proving a prime involves dividing it with every number smaller than it to ensure that they all result in a remainder. This exercise is fine for the first handful of primes tackled in a classroom, but larger number would require dozens, hundreds—potentially many billions—of divisions before being declared prime.

Eratosthenes, who invented a simple primality test, was the chief scholar at Alexandria's Great Library.

Ancient algorithm

High-speed super computers have taken over the search but the process they use to cross-check each candidate number were figured out long ago. Prime numbers lie at the heart of encryption algorithms that ensure security in telecommunications, but they were not lost on the ancients. They appear to have been used on the Ishango Bone, a 20,000-year-old carved calculating tool from the Congo.

In *Elements*, Euclid had shown that there is an infinity of primes, but that did not stop people looking for them. The Sieve of Eratosthenes is an early method for spotting primes attributed to a Greek mathematician and astronomer from the 3rd century BC. It is an algorithm—series of instructions—that reveal all the primes in a defined set of numbers, always starting at 2. First you remove all the numbers in the set that are a multiple of 2, the first prime. This will be all the even numbers; all other primes are odd. Next you strike out the multiples of 3, then 5 (4 has gone already), then 7. In a set of less than 100 numbers, that is sufficient to show all 25 primes. The process continues with ever larger primes as they appear. Despite being 2,300 years old, this is still the best way to find all the smaller primes, below the 10 million mark or so.

PRIMES VS PREDATORS

Periodical cicadas spend most of their live as wingless nymphs sucking sap from buried tree roots. To breed they must come above ground and molt into a winged adult form. Thousands of nymphs emerge at once, a prime feast for any waiting predators. However, the cicadas only emerge every 13 or 17 years. This prime-number period between each generation makes it impossible for predators to synchronize their life cycles with those of the cicadas.

A little theorem, big problem

Pierre de Fermat, an astounding 17th-century mathematician, made another tool for the search for primes. Fermat is best known for leaving the world in the dark over his "last theorem." Less well known is his "little theorem" on prime numbers.

Fermat's little theorem first saw light of day in a letter written in 1640 to Bernard de Bessy. Fermat announces that he has found that the result of $a^p - a$, where p is a prime and a is any number, will always be an exact multiple of p. True to form, Fermat does not show how he knows this; de Bessy is meant to figure it out for himself. Neither de Bessy, nor anyone else, could—and Fermat took the answer (if he had one at all) to his grave. Just under a century later, the Swiss mathematician Leonhard Euler proved the theorem—just one of a dazzling array of achievements.

The little theorem is used as a first step in the ongoing search for primes. The theorem is modified to $a^p-1 = b$, where b produces a remainder of 1 when divided by p. Values of a are chosen at random, and if they keep producing the correct answer, p becomes a "probable prime." However, just one product b that does not fit the theorem is enough to show that in this case p is a composite number.

A color-coded sieve showing the primes below 100. The colors equate to the composites of each prime from 2 upwards.

- ○ Multiples of 2
- ○ Multiples of 3
- ○ Multiples of 5
- ○ Multiples of 7
- ○ Prime numbers

2	**3**	4	**5**	6	**7**	8	9	10
11	12	**13**	14	15	16	**17**	18	**19**	20
21	22	**23**	24	25	26	27	28	**29**	30
31	32	33	34	35	36	**37**	38	39	40
41	42	**43**	44	45	46	**47**	48	49	50
51	52	**53**	54	55	56	57	58	**59**	60
61	62	63	64	65	66	**67**	68	69	70
71	72	**73**	74	75	76	77	78	**79**	80
81	82	**83**	84	85	86	87	88	**89**	90
91	92	93	94	95	96	**97**	98	99	100

14 Pi

THE RATIO OF THE CIRCUMFERENCE OF A CIRCLE TO ITS DIAMETER IS CALLED PI. That is the name for the Greek letter π, which is used as the number's symbol. With good reason π has been called the most famous number in the world.

For basic calculations, π is generally given as 3.14, but it took many years of hard work to say with any confidence that even this short approximation was valid. The complication lies in the fact that the number can be calculated to an infinite number of decimal places. No matter how hard we look, we will never find an exact figure for π.

Most likely π was discovered independently several times. The Babylonians certainly knew of it, and used a value of 3.125, which is fairly accurate, although we have no clear idea of how they arrived at it. In Egypt's Rhind Papyrus, Ahmes wrote, "Cut off 1/9 of a diameter and construct a square upon the remainder; this has the same area as the circle." This give an even better value for π of 3.1605.

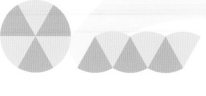

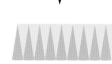

Classical calculations
The 5th-century-BC Greek mathematicians Antiphon and Bryson calculated π by inscribing one polygon just inside a circle and another

Archimedes calculated π by dividing a circle into a series of regular sectors and reorganizing them into a linear shape with curved edges. The length of this shape can be calculated using geometry and approximates to half of the circumference. The length gets ever closer to the true values as you increase the number of sectors, thus reducing the error introduced by the curved edges.

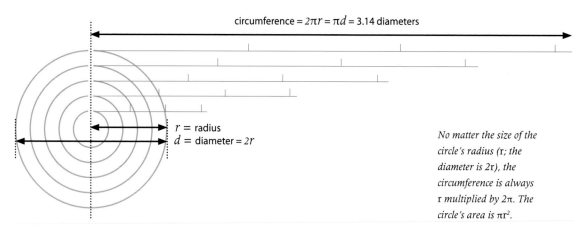

circumference $= 2\pi r = \pi d = 3.14$ diameters

r = radius
d = diameter = $2r$

No matter the size of the circle's radius (r; the diameter is 2r), the circumference is always r multiplied by 2π. The circle's area is πr^2.

PIEMS

A quirky literary phenomenon called piems are used to remember π. The number of letters in each word corresponds to a digit in the value of π

How   I   want   a   drink,   alcoholic slightly bigger one just outside. The area of the circle would lie somewhere between the areas of the two polygons. By increasing the number of sides on the polygons, closer and closer approximations to the area of the circle could be made.

Antiphon and Bryson were both tackling one of the great problems of classical geometry: is it possible to square a circle? In other words can you construct a square that was equal in area to a circle using only a ruler and a compass? For centuries, mathematicians struggled with this conundrum until 1882, when Carl Lindemann proved that π is a transcendental number. That means that not only is π composed of an infinitely long string of decimals, but that string of numbers contains no pattern that could be predicted. As a result, squaring a circle is against the rules of math.

The pyramids of Giza have a perimeter to height ratio which is a fair approximation of the value of π. Whether the architects based their calculations on π or simply liked this ratio for some other reason we can only speculate.

In the 3rd century BC, the great scientist and engineer Archimedes of Syracuse used a different approach. He used the perimeters of polygons to calculate the circle's circumference. He began with a hexagon and then doubled the sides four times to finish with 96-sided polygons inside and out of the circle. The end result of Archimedes' calculations was a value for π that lay somewhere between 3.140845 and 3.142857.

In later centuries this method was further refined. In the 3rd century AD, Chinese mathematician Liu Hui calculated π as a value of 3.141592104 using a polygon with 3,072 sides. In India 250 years later, Aryabhata reached 3.1416 using a 384-sided polygon.

Modern value

The computing power available today allows us to calculate π to extraordinary lengths. In 2011 Shigeru Kondo and Alexander Yee ran a custom-built computer for 191 days to calculate π to ten trillion decimal places. To put this into some sort of perspective, a mere 39-decimal value for π would be sufficient to calculate the diameter of the entire known universe so precisely that the degree of inaccuracy would be a tiny figure, even less than the radius of a hydrogen atom.

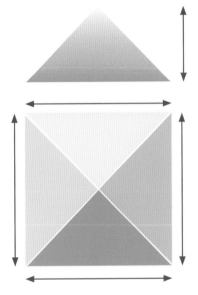

The ratio of perimeter to height of the Great Pyramid is 22:7 or 3.142.

of   course,   after   the   heavy   lectures   involving   quantum   mechanics!

15 Measuring the Earth

TODAY WE MIGHT USE A RADAR-MAPPING SATELLITE OR A POWERFUL TELESCOPE TO MEASURE THE SHAPE AND SIZE OF A PLANET OR STAR. In the ancient world, mathematicians made measurements of Earth with as little as a shadow cast by a pillar and trigonometry.

At the end of the 3rd century BC, a Greek mathematician devised a remarkably simple way of calculating the size of Earth that required him to take just one single measurement. Using deductive logic and the tools of Euclidean geometry, Eratosthenes, the chief librarian of Alexandria, measured Earth to an accuracy of almost 99 per cent! At the dawn of the classical Greek era in the 7th century BC, the world was thought to

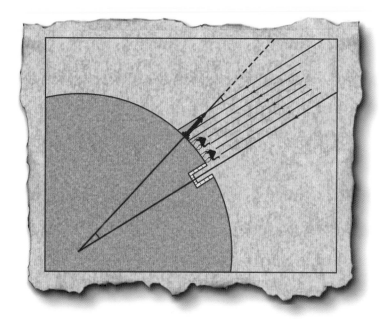

The well purportedly used by Eratosthenes still catches the sun at noon midsummer's day in Aswan. Eratosthenes realized that the angle of the noonday shadow at Alexandria was equal to the angle between Syene (ancient Aswan) and that city measured from the center of the planet.

be round but flattened into a disk. In 580 BC Anaximander of Miletus suggested it was actually a cylinder, with a flat landmass on top and a seething ocean curving around it. Sharp-eyed sailors had seen the curvature of the Earth first hand as approaching ships appeared over the horizon mast first, with the hull visible last of all. By the 5th century BC philosophers were suggesting that the planet was actually a sphere.

Shining idea

It was widely known that the Sun reached different altitudes—angles above the horizon—in different areas of the ancient world. When Eratosthenes heard that while the Sun cast shadows in Alexandria on midsummer's day, it was directly overhead at Syene (a city to the south, now known as Aswan) the master saw a way to calculate the circumference of Earth. He reasoned that the light from the Sun traveled in parallel beams. The beams over Syene arrived vertically, while they approached Alexandria at an angle, casting shadows. To calculate the angle of the sunlight in Alexandria Eratosthenes measured the length of a shadow cast by a pillar at noon on the summer solstice and geometry did the rest. The angle of the light measured from the vertical was the same as the angle between the two cities measured from the center of the Earth – 7° 12′, which is 1/50th of Earth's total circumference. The distance between the two cities was 5,000 stadia (an ancient Greek measure) and Eratosthenes rounded up his final figure to 252,000 stadia. The accuracy of that result depends on the length of his measurement. The standard Greek stadion of 185m produces an error of 16 percent. However, the local Egyptian stadion of 157.5m gives the result as 39,690km, less than 2 per cent out!

A century later, Hipparchus published the first trigonometry table to show how a triangle's angles relate to the lengths of its sides. He used it to measure the distances to the Moon and Sun. Trigonometry was developed further by Islamic scholars, and 1,100 years after Hipparchus, the Uzbeck scholar al-Biruni used trigonometry to calculate Earth's radius as 6,339.9 km, only 16.8 km under the modern value!

16 The Powers of Ten

PERHAPS UNSURPRISINGLY AS CIVILIZATION PROGRESSED AND STATES SWELLED INTO EMPIRES, the numbers that required recording also grew. After centuries of using unwieldy figures, the Chinese hit upon a simple fix.

The megas, gigas, and nanos that litter our conversations relate to the modern usage of powers of ten to describe the very large and very small.

The largest numeral available in ancient Greece was the *myriad* (10,000). A myriad myriad equated to 100 million, but a number a little more or less than this would have been laborious to write down and even more so to calculate. In 190 BC, Chinese mathematicians hit upon a way of harnessing the power of ten to simplify things. Any number could be written to a reasonable accuracy with a few digits and then multiplied by ten a set number of times to show its value. For example, 72×10^5 equates to 7,200,000, or 72 multiplied by 10 five times over. This process is termed *ten to the power five*, and larger powers make it possible to express numbers that are beyond our imagination. The technique for big numbers could be used to help with small ones too: dividing by 10 transformed numbers into tenths (10^{-1}), hundreths (10^{-2}) and so on. Decimal fractions had been born, but would take another 1,500 years at least to mature.

Prefix	Symbol	Multiplying factor	power
yotta	Y	1,000,000,000,000,000,000,000,000	10^{24}
zetta	Z	1,000,000,000,000,000,000,000	10^{21}
exa	E	1,000,000,000,000,000,000	10^{18}
peta	P	1,000,000,000,000,000	10^{15}
tera	T	1,000,000,000,000	10^{12}
giga	G	1,000,000,000	10^{9}
mega	M	1,000,000	10^{6}
kilo	k	1,000	10^{3}
hecto	h	100	10^{2}
deca	da	10	10^{1}
deci	d	0.1	10^{-1}
centi	c	0.01	10^{-2}
milli	m	0.001	10^{-3}
micro	μ	0.000,001	10^{-6}
nano	n	0.000,000,001	10^{-9}
pico	p	0.000,000,000,001	10^{-12}
femto	f	0.000,000,000,000,001	10^{-15}
atto	a	0.000,000,000,000,000,001	10^{-18}
zepto	z	0.000,000,000,000,000,000,001	10^{-21}
yocto	y	0.000,000,000,000,000,000,000,001	10^{-24}

17 The Modern Calendar

NATURE SUPPLIES THREE HANDY CYCLES TO MEASURE TIME—THE DAY, THE YEAR, AND THE LUNAR MONTH. The problem is that they don't conform with each other. As a result, creating calendars has always involved mathematical compromises and with that comes controversy.

Early uses of calendars included for agricultural planning, collecting taxes, and observing religious festivals. Ancient Egypt actually used two calendars at the same time. One calendar of exactly 365 days was used for administrative purposes, although since there are actually about 365 and a quarter days in the year, it gradually fell behind the seasons. There was also a moon-based religious calendar. Egyptian astronomers kept a close eye on the stars to follow the "real" seasons and adjust the calendars accordingly.

Julius Caesar was assassinated on the Ides of March (the middle of the month) in 44 BC. The ides of each month was traditionally associated with Mars the god of war, so it was an apt time for the attack. Without Caesar's own revisions to the calendar, he would have lived until the start of June.

Enter Caesar

The Romans liked to have things well organized, but by the first century BC their own calendar was looking a bit of a mess. Some features already look familiar to us: there were 12 months, starting with Ianuarius (January), while February was the shortest month. But the months still tried to follow the moon, and extra months were inserted rather haphazardly to catch up with the actual time.

When Julius Caesar became undisputed leader of Rome in 46 BC, he was determined to sort the calendar out. He may have got the idea while campaigning in Egypt the previous year. Egypt at that time was ruled by the Greek successors of Alexander the Great, and the Greeks had long known that the actual year was about 365 and a quarter days. (Back in 238 BC, Egypt's Greek ruler had made the first attempt to approximate this figure, by proposing leap years of 366 days every four years—but the locals had so far resisted this innovation.)

Caesar, with advice from astronomer Sosigenes of Alexandria, revived the leap year idea, and also adjusted the lengths of the months. By this time the old Roman calendar was running three months ahead of itself, and Caesar also wanted January pulled back to its position after midwinter. So he decreed that the year now called 46 BC would have two extra months, making it 445

Christopher Clavius, a German mathematician, was the chief architect of the Gregorian calendar, even though he was a staunch geocentrist, insisting that the Sun went around the Earth, not the other way around.

Sunlight shining on a 44-meter brass meridian installed in the Church of Santa Maria degli Angeli in Rome in 1702 shows the date of the vernal equinox from which Easter is calculated (as the first Sunday after the next Full Moon).

days long, bringing everything back into line. The reforms survived Caesar's murder two years later, although he died before the first leap year and officials mistakenly started adding a leap year every three years instead of four. But Caesar's successor Augustus put things straight, and the new "Julian calendar" ran smoothly in Europe for the next 1,600 years.

The Gregorian calendar

Caesar may well have known that in the long term his new calendar would eventually get out of step with the seasons. Already ancient astronomers could measure the length of the year very exactly by observing a once-a-year position of the Sun called the vernal equinox. Such observations showed the average year to be 11 minutes shorter than 365.25 days, meaning that the Julian calendar gets one day behind every 128 years.

By late medieval times in Europe, the Julian calendar had slipped over a week—especially a nuisance for churchmen needing to work out the date of Easter. There were various attempts to get the Pope to institute a change. Eventually, in 1578, Pope Gregory XIII decided to act. He took expert advice and announced a small but effective tweak to the Julian system of leap years, so that century years (1600, 1700, etc) would not be leap years unless they were also divisible by 400.

Like Caesar, Gregory wanted to pull the dates of the year back to an earlier position so Easter occurred in spring each year. He announced that in the year of implementation, 1582, 10 days would be left out of October, with the 4th being followed immediately by the 15th. (An 11th was later removed as well.) This change was the most significant to ordinary people, although reports of rioting over the loss of time appear to be apocryphal. Gregory's simple adjustments made the calendar very accurate. The Gregorian year will not fall one day behind until the year 3719.

ADOPTING THE CALENDAR

As the head of the Catholic Church, Pope Gregory's reforms happened first in Catholic countries, such as Spain and Poland. Protestant countries were slower to make the changes. Sweden opted to remove 11 days gradually over 40 years, meaning it was using a unique dating system for all that time. The British Empire (including America) made the changes in 1752. The dates of the financial year were not adjusted and even today Britain's companies pay taxes based on a year beginning on April 6. Meanwhile Turkey carried on using the Julian calendar until 1929.

William Hogarth's satirical painting An Election Entertainment, pokes fun at British politicians in the mid 1750s. Bottom right a placard reads "Give us our eleven days" referring to the way the two parties, the Whigs and the Tories, even argued over the date.

18 Diophantine Equations

IN THE 3RD CENTURY AD, DIOPHANTUS OF ALEXANDRIA PUBLISHED *ARITHMETICA,* which translates as "the science of numbers," and this ancient work became a major milestone in number theory—the study of integers.

The *Arithmetica* contained 130 equations which have come to be known as *Diophantine*. A Diophantine equation is one where the only permitted variables are integers—whole numbers. In modern terms they are a small group of a wider set of polynomial, or algebraic, equations. (A polynomial is an expression with two or more algebraic terms made up of at least one unknown variable, normally *x*.) Diophantus has been called "the father of algebra," although that word, its modern letter-based notations, and many of concepts associated with it were several centuries away yet.

An example of a Diophantine equation asks something like this: "A father is 1 year younger than twice the age of his son. The digits in his age, represented as AB, are reversed for his son's age. How old can the father and son be?" The only possible answer is that the father is 73 and the son is 37. In many cases, this is arrived at by trial and error, though after the solution is reached, a mathematical proof can be put together in hindsight.

Diophantus was careful to consider only the problems that he thought to be solvable with a single answer. He always looked for positive solutions—negative-number answers were not considered valid. As such, Diophantine equations were sometimes dismissed as puzzles, but many remained unsolved for centuries. Pierre de Fermat, the proposer of the now famous "Last Theorem," was studying these very puzzles in 1637, when he thought of a Diophantine equation that had no solution. He noted it in the margin of page 85 of his copy of *Arithmetica*: "If integer *n* is greater than 2, then there are no three integers where $x^n + y^n = z^n$." Fermat left no proof and it took until 1994 to find one.

The original work of Diophantus had Arithmetica *in 13 volumes but only six have survived. This is a Latin translation published in 1621.*

DIOPHANTUS RIDDLE

Even the inscription on Diophantus' tomb was a polynomial expression—with the unknown value being the mathematician's age. This is a rough translation: "Passerby, this is the tomb of Diophantus. It is he who with this surprising inscription tells you how many years he lived. Diophantus' youth lasted one sixth of his life. He grew a beard after one twelfth more. After one seventh more of his life, he married. Five years later, he and his wife had a son who, once he reached half the age of his father, suffered an unfortunate death. His father had to survive him, crying, for four years. From this, deduce his age." (Answer: 84)

19 Hindu-Arabic Number System

THE DECIMAL NUMBER SYSTEM USED ACROSS THE WORLD TODAY HAS ITS ROOTS IN 6TH-CENTURY INDIA. However practical the familiar digits of zero through nine may appear to the modern mind, it took another millennium for them to be accepted as the world's numbers.

Brahmi		—	=	≡	+	𑁫	𑁬	𑁭	𑁮	𑁯
Hindu	०	१	२	३	४	५	६	७	८	९
Arabic	٠	١	٢	٣	٤	٥	٦	٧	٨	٩
Medieval	o	ɪ	2	3	ꝝ	ϥ	6	ʌ	8	9
Modern	0	1	2	3	4	5	6	7	8	9

The evolution of our modern digits can be traced from Brahmi numerals dating back to the 3rd century BC. The zero appeared around the same time as a placeholder in the Babylonian base-60 counting system, but made a seamless addition to the Hindu-Arabic numbers.

While the western world was slogging away with Roman numerals—hard to write, read, and calculate—India and China had always benefited from a positional system similar to our modern one. The difference was that 10s, 20s, and other tens had their own symbols. The 6th-century innovation was to add a zero and to do away with numerals higher than nine. This system spread rapidly with the Islamic conquests of the 7th century, and was soon in use from Cordoba to Calicut. As a result we know it today as the Hindu-Arabic number system.

The Hindu-Arabic system was the source of heated debate for centuries in Europe between the algorists, its supporters, and the abacists. The abacists insisted Roman numerals and a counting board were superior to the written calculations of the algorists. The debate finally fizzled out in the 16th century when Roman numerals were confined to history.

Liber Abaci

Nevertheless, even by the 12th century, Europeans were still counting and calculating in Roman numerals. Merchants who ventured across the Mediterranean to trade in Africa were astonished by the speed with which their Arab colleagues could calculate and then present the latest prices. One such traveler was Leonardo of Pisa, a young Italian now better remembered as Fibonacci. Leonardo accompanied his father on trips to Arab lands, where he was not only won over by the number system but also learned from the local scholars how to put it into mathematical action. In 1202, the Italian presented his findings in a book, *Liber Abaci*. That title translates as Book of the Abacus, but it was anything but. Leonardo presented the new number system and the rapid calculation methods it allowed as practical tools for merchants. In so doing he dragged European math out of the Dark Ages and ignited a fire in commercial culture that would that put Europe at the heart of world power for centuries to come.

20 Algorithms

AN ALGORITHM IS A STEP-BY-STEP PROCEDURE FOR ANSWERING QUESTIONS THAT SEEKS TO ELIMINATE TRIAL AND ERROR IN THE INVESTIGATIVE PROCESS. The word is derived from Arabic, with great breakthroughs made by Islamic scholars in the 9th century. However, the concept was being practiced more than a millennium before.

Al-Khwarizmi has a long list of accolades. He was instrumental in the introduction of the decimal point to European math, he was a pioneer of trigonometry, and he made some of the most accurate maps of the age.

The word *algorithm* is often associated with computers today, perhaps taking a central role in an action movie, where a high-tech computer algorithm is poised to turn the world order on its head. It may then be a little deflating to hear that an algorithm is just a set of instructions that when carried out in a specific order will always produce a desired result. At their simplest all computer programs are algorithms. Euclid himself formulated an algorithm 2,300 years ago for his seminal work *Elements*. However, he would not have called it an algorithm. That term derives from *Algoritmi*, the Latinized name of Muhammad ibn Musa al-Khwarizmi, a Persian mathematician from the 9th century AD.

EUCLID'S ALGORITHM

The process now known as Euclid's algorithm is for finding the greatest common denominator (GCD) of two numbers, in other words a third number that is the largest value that can be divided exactly into both of them. The starting numbers have to be composites (non-primes; primes have not common divisors). The process starts by dividing the smaller into the larger, and presenting the result as the quotient (the number of times it divides) and the remainder. Then the remainder is divided into the smaller starting number, followed by the resulting remainders into their predecessors, until the remainder is zero.

GCD OF 4,433 AND 1,122:

4,433 IS DIVIDED BY 1,122: QUOTIENT 3, REMAINDER 1,067

1,122 IS DIVIDED BY 1,067: QUOTIENT 1, REMAINDER 55

1,067 IS DIVIDED BY 55: QUOTIENT 19, REMAINDER 22

55 IS DIVIDED BY 22: QUOTIENT 2, REMAINDER 11

22 IS DIVIDED BY 11: QUOTIENT 2, REMAINDER 0

THE GCD OF 4,433 AND 1,122 IS 11

Formal process

Al-Khwarizmi is linked to algorithms through his book *Compendious Book on Calculation by Completion and Balancing*. In this he set down a standard process for solving linear and quadratic equations (also introducing many of the concepts of *algebra* in the process). He developed a formal procedure that worked for any problem. This involved reducing it to one of six standard forms, and then removing negative units, roots, and squares.

Taking the trial-and-error out of solving mathematics, algorithms have became tools for automating reason. In the 1840s, Ada Lovelace developed an algorithmic process for use in Charles Babbage's primitive computer (should it have been built); while 1,100 years after al-Khwarizmi, the algorithm was central to the Turing Machine—a thought experiment that was the forerunner of digital computing.

21 Cryptography

CODES ARE AS OLD AS SECRETS, AND TECHNIQUES FOR BREAKING THEM FOLLOWED HOT ON THEIR HEELS. At first codes had no need for math. However, in the 9th century, an Arab philosopher developed a mathematical method for revealing the hidden meaning.

Strictly speaking a code is a way of hiding meaning by changing whole words. So often a single "code word" is enough to pass on a message. For example everyone knows what is meant by *D-Day* today, but mercifully few knew the code prior to the event. What might be mistermed a code, a string of meaningless letters, numbers, or other characters is actually a cipher, which has been encrypted using a key. The key is the system that converts the original message—the *plaintext* in cryptographic jargon—into something unreadable.

The key is the weak point in keeping the message secure. Without it code-breakers have to use "brute force" by just trying every possible permutation of letters. Until the innovations of the 9th century, this would have probably worked despite taking a very long time. Until this point a substitution key was sufficient, in which the letters in the plaintext were substituted according to a prearranged scheme.

Al-Kindi, a polymath from Baghdad, put a stop to all that by developing frequency analysis. This relies on the fact that certain letters appear more frequently than others. The chances are that the most common character in an encrypted English phrase is an *e*, and so a prospective key is built around that. If that does not work, then *t*, the next most common letter is used. The longer the cipher, the more chance that it conforms to the average. Even the most cunning substitution keys were meaningless in the face of al-Kindi's system. A notable victim was Mary Queen of Scots who encrypted her letters with a set of unique characters. Frequency analysis revealed evidence of treason within and she was executed by Elizabeth I in 1587.

A 17th-century linguistics table shows characters used in the world's main alphabets. The 1679 work called Turris Babel *(Tower of Babel) by the German Athanasius Kircher also contained details on their use in cryptography.*

IGKYGX IUJK

CAESAR CODE

A simple cipher is named after Julius Caesar, who used it to encrypt his orders. The key below should make it easy enough to figure out what our encrypted title says. This particular cipher has a key of 7—the alphabet has been shifted 7 places. Yet even if Caesar's top brass frequently altered their keys this code was far from secure. Nevertheless it would have foiled the barely literate non-Latin speaking enemy that the Romans frequently joined in battle.

A	B	C	D	E	F	G	H	I	J	K	L	M	N	O	P	Q	R	S	T	U	V	W	X	Y	Z
G	H	I	J	K	L	M	N	O	P	Q	R	S	T	U	V	W	X	Y	Z	A	B	C	D	E	F

22 Algebra

NO SCIENCE CARTOON IS COMPLETE WITHOUT A BLACKBOARD SCRAWLED FULL OF COMPLEX-LOOKING EQUATIONS. But rather than a shortcut to abstraction, algebra is a grounding principle – the grammar that underpins the language of mathematics.

A
TREATISE
OF
ALGEBRA,
ᴮᴼᵀᴴ
Hiſtorical and **Practical.**
SHEWING,
The Original, Progreſs, and Advancement thereof, from time to time; and by what Steps it hath attained to the Heighth at which now it is.

With ſome Additional TREATISES,

I. Of the *Cono-Cuneus*; being a Body repreſenting in part a *Conus*, in part a *Cuneus*.
II. Of *Angular Sections*; and other things relating thereunto, and to Trigonometry.
III. Of the *Angle of Contact*; with other things appertaining to the Compoſition of Magnitudes, the Inceptives of Magnitudes, and the Compoſition of Motions, with the Reſults thereof.
IV. Of *Combinations, Alternations,* and *Aliquot Parts.*

By J O H N W A L L I S, D. D. Profeſſor of Geometry *in the Univerſity* of Oxford; *and a Member of the* Royal Society, London

LONDON,
Printed by *John Playford*, for *Richard Davis*, Bookſeller, in the Univerſity of O x f o r d, M. DC. LXXXV.

John Wallis's Treatise of Algebra *from 1685 was both a practical guide to the methods of algebra of the time, and also a guide to the history of algebra.*

The term *algebra* was another bequest from Muhammad ibn Musa al-Khwarizmi, an Islamic scholar who was the director at the magnificently named House of Wisdom (Bayt al-Hikma) in Baghdad. This unrivalled intellectual hothouse, the members of which excelled in mathematics, astronomy, alchemy, medicine, astrology, zoology, and geography, popularized the use of Hindu numerals and the Indian decimal point. In a seminal text of 820 AD, al-Khwarizmi introduced the word *al-Jabr*, meaning "restoration" or "completion," from which our term algebra derives. Although he was not the first to represent unknown quantities as non-number values, al-Khwarizmi's contribution was to formalize the way of restoring balance to equations. Four centuries later in the early 13th century, Italian mathematician Leonardo of Pisa, or Fibonacci, introduced algebra to a Europe still struggling with Roman numerals.

What algebra is for

Algebra allows you to express "word problems" in mathematical language. Al-Khwarizmi used it to solve quadratic equations (where a variable is squared). Algebraic expressions such as $2x - 3 = 5$ are called equations, as opposed to identities ($2 + 2 = 4$) and relationships ($F = ma$). The x and y notations we are used to today came much later; al-Khwarizmi just used words to present problems. However, he balanced expressions: whatever operation was performed on one side was also done to the other

The power of algebra extends much further than figuring out unknown quantities. Substituting symbols for numbers takes a mathematical expression beyond the specific to the general. Equations are universal and are independent of the numbers inputted—they explain why a solution always works.

In the 1600s, the French mathematician René Descartes linked algebra to geometry, allowing equations to be plotted as shapes on a graph. Similarly, functions of x (operations that process values according to a rule) could be mapped onto axes and analyzed, leading the way to describing geometries of higher dimensions.

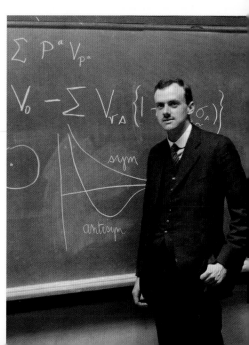

Paul Dirac, a British quantum physicist, used algebra to describe the properties of an electron, and in so doing revealed the existance of antimatter—all from an algebraic formula.

23 Fibonacci Sequence

THE NAME LEONARDO OF PISA IS NOT ONE THAT IS HERALDED DOWN THE AGES, ALTHOUGH HIS NICKNAME FIBONACCI IS MUCH MORE FAMILIAR. The impact of this 13th-century Italian is impossible to quantify. Not only did he transform the way Europeans count, he also unleashed one of the most amazing number sequences in mathematical history.

Hidden away in Fibonacci's *Liber Abaci* is a seemingly mundane problem about how livestock breeders can predict stock numbers. However, the so-called Rabbit Problem revealed a pattern of numbers that would appear time and again in the mathematics of growth, proportion, and even beauty: *"How many pairs of rabbits will we have a year from now if we begin with one pair that produces another pair each month, which in turn become productive after two months?"* On the face of it, Fibonacci's problem would not look out of place in a high-school test, so what's the answer: begin with one pair; one month later you still have one pair, although the female is now pregnant; two pairs by the second month; three in the third (remember the newborns can't breed just yet); then five pairs in the fourth month, eight in the fifth. By the end of the year you have 144 rabbits. Closer scrutiny of the sequence reveals that each number is the sum of its two previous ones. This list of numbers, now known as the Fibonacci sequence, has been found at play in the shape of flowers and in other patterns in nature and in art.

The seeds in a sunflower spiral outward clockwise and counterclockwise. The number of either is always found in the Fibonacci sequence. The number of the clockwise spirals will always be the next number in the sequence below or above the number of counterclockwise spirals.

GOLDEN CONNECTION

The Fibonacci sequence has a startling connection with the golden ratio, expressed as the number phi, which is arrived at in an entirely different way. However, if you divide successive Fibonacci numbers by their predecessors, the resulting quotients tend towards phi. They never get there precisely but by the tenth place in the sequence the results are less than a thousandth out—and getting closer every time.

Place	Number	Fibonacci number/ predecessor	Difference from phi (Φ)
1	1		
2	1	1.000000000000000	-0.618033988749895
3	2	2.000000000000000	+0.381966011250105
4	3	1.500000000000000	-0.118033988749895
5	5	1.666666666666667	+0.048632677916772
6	8	1.600000000000000	-0.018033988749895
7	13	1.625000000000000	+0.006966011250105
8	21	1.615384615384615	-0.002649373365279
9	34	1.619047619047619	+0.001013630297724
10	55	1.617647058823529	-0.000386929926365
11	89	1.618181818181818	+0.000147829431923
12	144	1.617977528089888	-0.000056460660007
13	233	1.618055555555556	+0.000021566805661
14	377	1.618025751072961	-0.000008237676933
15	610	1.618037135278515	+0.000003146528620
16	987	1.618032786885246	-0.000001201864649
17	1597	1.618034447821682	+0.000000459071787
18	2584	1.618033813400125	-0.000000175349770
19	4181	1.618034055727554	+0.000000066977659
20	6765	1.618033963166707	-0.000000025583188

24 Perspective Geometry

GEOMETRIC OR LINEAR PERSPECTIVE IS A MEANS OF CREATING THE ILLUSION OF DEPTH IN A PICTURE. The techniques for producing realistic representations of three-dimensional objects on a two-dimensional surface were first formulated in Italy in the 15th century.

Linear perspective makes use of the fact that objects appear to get smaller the farther away they are, and that parallel lines and planes extending away from the viewer converge at a vanishing point. The 13th-century painter Giotto frequently created the impression of depth by using sloping lines. Those above the eye line of the observer sloped down, while those below sloped up. Lines to the side inclined

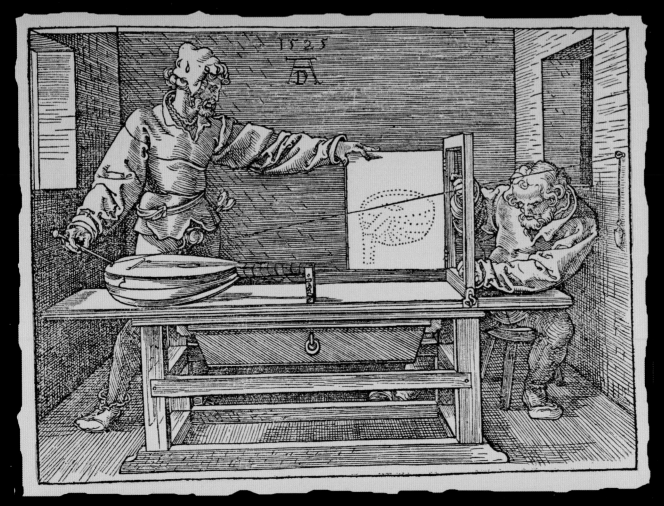

Albrecht Dürer presented a compendium of geometric techniques for producing perspective in art in his Underweysung der Messung mit dem Zirckel und Richtscheyt *(Four Books on Measurement). This woodcut from this 1522 work gives instructions on how to draw a lute accurately.*

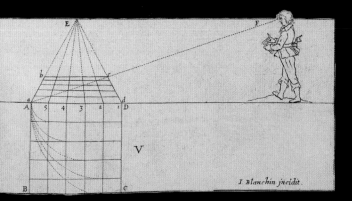

This diagram by the 17th-century French monk, painter, and mathematician Jean-Francois Niceron, shows how parallel lines merge to a vanishing point when drawn in perspective.

towards the center. The first person credited with a mathematical understanding of perspective is the Italian Filippo Brunelleschi. Brunelleschi was an architect (Florence Cathedral was one of his works) and he worked out the relationship between the actual length of an object and the way its length appeared to change depending on how far away it was from the observer.

Master works

Fellow Italian Battista Alberti was the first to write down his ideas about linear perspective in his 1435 book *Della Pittura* (On Painting). Alberti developed the idea of the visual pyramid. The pyramid has its point of origin, or apex, in the eye of the viewer. The sides of the pyramid extend outwards from the apex following the edge of the field of vision. The painting could be imagined as a plane that intersects the visual pyramid and the apex of the pyramid as being at the ideal point for viewing the image. The vanishing point in the image, on which lines converged, was envisaged as being as far beyond the plane of the painting as the apex was in front of it. Artists had to see a painting as a window through which the observer sees the scene. A horizon runs across the canvas at eye level, and the vanishing point is located somewhere near the center of this line. This technique is called one-point perspective.

The most mathematical account of perspective during the 15th-century Renaissance was from Piero della Francesca, a leading mathematician as well as one of the greatest artists of his day. He developed mathematical formulae to compute the size an object should be painted on the canvas relative to its distance from the observer. He also dealt with depicting more complex objects using a method involving two rulers, one to measure width, the other height, in effect developing a coordinate system against which to plot the correct position of points on the object being represented.

CHANGE OF PERSPECTIVE

Hans Holbein was a master of the art of perspective. His famous painting *The Ambassadors* (1533) shows two men together with a number of objects associated with arts and learning—a consummate demonstration of his ability to render objects and people in a realistic manner. However, in the foreground of the picture there is a distorted object. It is only when the viewer observes it from the far right of the painting that it is revealed as a human skull. Holbein has used anamorphosis, skewing the perspective so that the object only appears correctly from a particular viewpoint.

25 Non-Linear Equations

A LINEAR EQUATION PRODUCES A STRAIGHT LINE WHEN PLOTTED ON A GRAPH. In the 1580s, the musician father of Galileo is credited with discovering the first non-linear relationship.

The relationship between musical tones and the length of a plucked string as discovered by Pythagoras is a linear one. For centuries it was assumed that increasing the tension on a string to produce higher pitched notes followed the same relationship. However, Vincenzo Galilei demonstrated that the ratio of an interval (a step up or down in pitch) was indeed proportional to the length of the string: it varied according to the square of the string's tension. For a wind instrument, such as a flute, the interval varied as the cube of the volume of air vibrating inside. So an interval of a perfect 5th could be produced by similar strings differing in length by a ratio of 3:2, by similar strings differing in tension by 9:4, and by columns of air differing in volume by a 27:8 ratio.

Vincenzo Galilei was a composer and lute player. While his older son became a scientist, his younger one, Michelagnolo was a virtuoso musician.

26 Pendulum Law

A swinging, suspended lamp in Pisa Cathedral is still named for Galileo.

GALILEO GALILEI IS SAID TO HAVE DISCOVERED THE PRINCIPLE OF THE PENDULUM AS A STUDENT IN 1582 while at mass in the cathedral of Pisa, Italy. According to the story, he noticed that a lamp hanging from the ceiling was swinging back and forth.

The lamp had been set in motion by the verger who had just lit its candles. Galileo timed the swing of the lamp, using his own pulse as a measure, and realized that, though the extent, or amplitude, of the swing diminished each time, each complete back and forth oscillation of the lamp always took the same amount of time.

Whether this story is true or not, Galileo began a comprehensive study of pendulums some years later, around 1602. He was also able to show that the period of oscillation of a simple pendulum is proportional to the square root of its length. Changing the mass of the pendulum bob has no effect on the period—a heavier bob oscillates with the same frequency as a lighter one. The discovery of this property of pendulums, called isochronism (meaning

equal time) was to lead to the invention of the first accurate mechanical clocks. Galileo himself actually produced a design for a pendulum clock but it was never constructed.

Grandfather of clocks

Dutch scientist Christiaan Huygens built the first pendulum clock in 1656. Mechanical clocks had been around since the early 14th century—the earliest we know of was one in Milan in 1335. They operated by means of a weight falling at a controlled rate of descent. Such clocks gained or lost 15 minutes or more in the course of a day. The clock Huygens designed used the same mechanism of the old mechanical ones to turn the hands but used the swing of a pendulum to release it. Moving the bob up or down regulated the period of the swing to exactly one second.

Huygens' work on pendulums was used by Robert Hooke and others to formulate a mathematical basis for oscillations—even the vibrations of atoms.

Even though his clock was an impressive timekeeper, losing just a few seconds each day, Huygens continued to look for improvements. He used math to analyze the motion of pendulums, showing that if a pendulum's swings were all to be exactly isochronous irrespective of starting position, the bob must follow an arch-shaped cycloid curve. However, pendulums swing in circular arcs, so the time taken for each swing changes slightly with the angle of swing. In the case of his clocks, a small swing angle means any variations in period are negligible, but they lack the power to operate the clockwork. In 1673, Huygens built a clock where the top of the pendulum was made of flexible wire that swung against a curved metal surface, which changed the length of the pendulum slightly as it swung, correcting any differences in period due to variation in angle. This allowed large swing angles that could turn the clockwork.

A replica of the pendulum clock designed by Galileo in the 17th century.

In 1851, Léon Foucault erected a large pendulum in Paris to provide the first proof of Earth's rotation that did not rely on the motion of heavenly bodies. The pendulum maintained its direction of swing, while the planet rotated beneath—so over several hours the pendulum appeared to shift around its circular arena.

27 *x* and *y*

FRANCOIS VIÈTE IS SOMETIMES CALLED THE "FATHER OF ALGEBRA," although he did not invent it. But without him, those familiar equations would be unrecognizable.

The contribution of the 16th-century French mathematician was to introduce the use of letters to represent known constants and unknown quantities in equations—he used consonants for known quantities and vowels for unknowns. Viète's approach to algebra allowed him to solve a number of problems that had defeated earlier scholars. It also allowed mathematicians to analyze the relationships between the values of the coefficients of the original equation and its solutions.

Viète succeeded in cracking a code of more than 500 characters used by King Philip II of Spain during his war with the French Huguenots. Philip was so certain that his code was secure that once he had discovered that the French were aware of his plans he complained to the Pope that satanic black magic must have been used against him.

The first equals sign (=) appeared in the Whetstone of Witte, a 1559 math book by Englishman Robert Recorde.

28 Ellipses

AN ELLIPSE, A TYPE OF CURVE THAT IS LIKE A FLATTENED CIRCLE, WAS WELL KNOWN TO THE MATHEMATICIANS OF ANCIENT GREECE as the shapes produced by slicing up cones, a complex but manageable geometric task. In the 17th century, the math behind the humble ellipse was used to turn our understanding of Earth and the Universe on its head.

For centuries people had believed that the Earth was at the center of the Universe and all the stars and planets were set in spheres that made music as they turned around the Earth. Around 260 BC the Greek astronomer Aristarchus had suggested that the Earth orbited the Sun, but this notion was much too far-fetched for the people of the time. It was some 1800 years later, in 1543, that Polish astronomer Nicolaus Copernicus shook astronomy awake when he showed that Aristarchus had been right all along. The easiest way to explain the observed motion of the planets was to have them moving around the Sun, and Earth was just one of six planets circling the star.

The German Johannes Kepler took things a step further when he made a careful study of the work of his mentor, the Danish astronomer Tycho Brahe. Brahe had made some detailed observations of the movements of the planet Mars. Six years of lengthy calculations led Kepler to the conclusion that the observable data just didn't fit with the idea that Mars *circled* the Sun at all. The only thing that made sense was that Mars was moving in an ellipse.

Change of focus

A circle has just one internal focal point or focus, the center. The flattened circle that is an ellipse, however, has two internal foci, equidistant from the center point along the major or long axis of the shape. In an ellipse, the sum of the distances from any point on the curve to the foci inside the curve is a constant. An ellipse becomes more eccentric (less circular) as the distance between the foci increases. Circles and ellipses belong to the family of curves known as the conic sections, formed by a plane cutting through a cone.

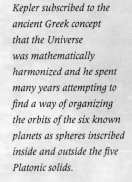

Kepler's elliptical results lead to the statement of his first law of planetary motion in 1609 (coincidentally the year that Galileo turned his telescope spaceward and discovered the moons of Jupiter). Kepler's work is sometimes referred to as the law of ellipses; it states that all planets orbit the Sun in a path that forms an ellipse with the Sun positioned at one focus of the ellipse.

Kepler subscribed to the ancient Greek concept that the Universe was mathematically harmonized and he spent many years attempting to find a way of organizing the orbits of the six known planets as spheres inscribed inside and outside the five Platonic solids.

Ellipses form part of the "conic sections," curves formed by angled planes slicing through cones. As well as ellipses (and circles), conic sections produce parabola curves (left) and hyperbola (right).

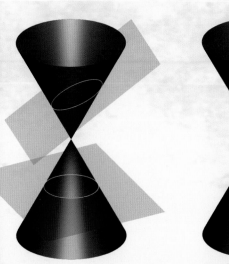
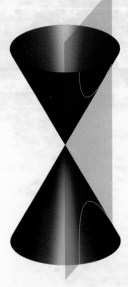

Driving force

Kepler puzzled over what was pushing the planets around. He theorized that there was some sort of magnetic force in play that attracted the planet towards the Sun in one half of its orbit and repelled it in the other half. Kepler knew that the gravitational attraction of the Moon caused the tides, and yet he failed to connect it to the orbits of the planets. He missed this because he believed that there must be a constant pushing force that kept the planets moving around their orbits. The explanation as to why the planets moved in ellipses and not circles would have to wait for another 80 years and Isaac Newton's discoveries concerning gravity and motion.

29 Logarithms

BEFORE CALCULATORS AND COMPUTERS, LOGARITHMS WERE THE BEST WAY TO CARRY OUT CALCULATIONS. They were invented by John Napier, Laird of Merchiston, an eccentric Scottish aristocrat seldom seen in public without his pet rooster and spider.

Logarithms are a crafty way to simplify multiplication by changing it into addition. They do this by making use of a striking mathematical fact about multiplying powers of numbers—to multiply them together, just add the powers. For example, take 2^2 multiplied by 2^3: $(2\times2) \times (2\times2\times2)$ = $2\times2\times2\times2\times2 = 2^5$. The result is 25, because we have multiplied 2 by itself $3 + 2 = 5$ times. Similarly with 10s: $100 \times 1000 = 10^2 \times 10^3 = 10^5 = 100,000$

John Napier had the idea of expressing all numbers as powers of another number, by using powers that were not whole numbers. This sounds confusing, but a square root can be expressed as a number to the power 1/2 or 0.5: for example, $5^2 = 25^1$, therefore $25^{0.5} = 5$. He published his ideas in his modestly titled 1614 work, *Mirifici Logarithmorum Canonis Descriptio* (Description of the Marvelous Canon of Logarithms), coining the term *logarithm* from the Greek words "logos" (reason) and "arithmos" (number).

The title page of Napier's Mirifici Logarithmorum Canonis Descriptio *(Description of the Marvelous Canon of Logarithms) published in 1614.*

Using base 10

Napier's logarithms were rather cumbersome, as they were based on powers of the number $1–10^{-7}$, or 0.9999999. His collaborator, the eminent English mathematician Henry Briggs, suggested a version of logarithms using the number 10 as the base. In this system, the logarithm of a number is the power you would need to raise 10 to get that number. For example, the log to the base 10 of 100 is 2 (written $\log_{10}100 = 2$), because $10^2 = 100$.

Briggs also developed the first tables of logarithms, published in 1624. These became a fantastic labor-saving device, used by scientists and engineers for some 250 years until the

This graph shows how doubling (multiplying by 2) a normal value results in a $\log_2$ increase of just one.

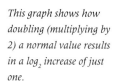

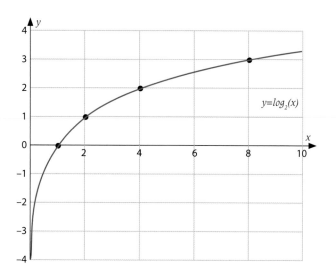

$y=log_2(x)$

A chart of linear values (top) and their $\log_2$ equivalents (bottom).

1	2	4	8	16	32	64	128	256
0	1	2	3	4	5	6	7	8

advent of pocket calculators in the 1970s. The other great calculating tool used over much of this same period, the slide rule, is also based on logarithms. To multiply any numbers together using log tables, you just need to find their logarithms from the table and add these up. Looking up the resulting number in a table of *antilogarithms* gives you the answer to your sum.

Mathematically, logarithms are simply the reverse of exponentials (powers of a number) and can be to any base. For example, $\log_2 8 = 3$, because $2^3 = 8$. Today, in mathematics, *natural* logarithms, which use the number *e* as the base, are most often used to describe the way quantities are seen to change in nature. Similar processes are seen in economic data, which is where *e* originally appeared to mathematicians.

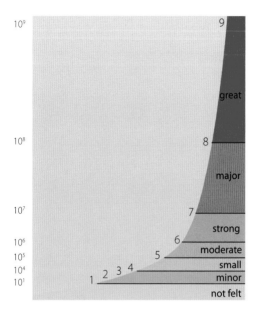

Logarithms at large

One reason for using log scales is that they are a much more compact and clear way to represent datasets with large ranges, compared to a normal linear scale such as the ones we use for temperature. For example, the pH scale, which measures the acidity and alkalinity of solutions, is based on the log of the concentration of hydrogen ions in a solution. In reality, such concentrations vary from around 1.0 mole for a concentrated strong acid, to around 0.00000000000001 mole for a strong alkali—a factor of some 100,000,000,000,000. On the pH scale, this is compacted to a range of between 0–14. (A mole by the way is a unit of quantity, while a dozen is 12, a gross 144, a mole is 6.0221415×10^{23}.)

One of the most familiar examples of a log scale is the Richter scale, used for measuring the size of earthquakes. Developed in 1935 by Charles Richter in California, the scale is devised so that an increase of one unit represents a tenfold increase in the power of the earthquake. So an earthquake of magnitude 5.0 on the Richter scale carries ten times the energy of one measuring 4.0, and 100 times the size of a 3.0-magnitude earthquake.

THE DECIMAL POINT

The inventor of logarithms, John Napier also had a role in making the decimal point such a familiar device. Throughout his posthumously published work of 1619, *Description of the Marvelous Canon of Logarithms*, Napier makes use of the modern notation of a dot to separate off the part of a decimal number that is less than one. The idea of decimal calculations had been suggested earlier, by the Dutch mathematician Simon Stevin in 1585. Napier's notation was simpler and easier to use than Stevin's and later became the established form, although a comma rather than a dot is now used in most European countries.

$$184.54290$$
$$184⓪5①4②2③9④0$$

Two decimal notations: The modern one at the top hails from John Napier, while the lower one was developed by Stevin. The circled numbers relate to the negative power of ten of the number proceeding them.

Psychologists have also discovered that it may be natural for humans to think of numbers themselves as forming a log-like scale. People brought up on school mathematics are taught to visualize numbers spaced equally on a number line, like the units on a tape measure. But studies with people of an isolated Amazonian tribe, the Mundurucu, reveal that even adults among them visualize numbers as positioned more closely as they get larger, producing an approximately logarithmic scale. Here, for the numbers 1 through 10, 1 and 2 are furthest apart while 9 and 10 are closest together— just as preschool children tend to do in more developed societies. Logarithms seem to reflect our perceptions more closely than counting numbers, from the diminished enjoyment of the fifth chocolate relative to the first, to the apparent quickening of time as we get older.

30 Napier's Bones

JOHN NAPIER WAS KNOWN FOR A WIDE RANGE OF INNOVATIONS, and today one of Edinburgh's universities is named for him. His celebrated invention was a simple calculator.

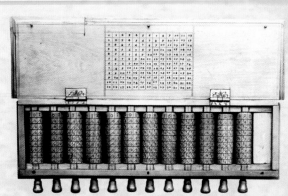

The device is known as Napier's bones. It is a set of ten rods that help the user to carry out long multiplication sums and other arithmetic operations, such as division and even finding square and cube roots. The faces of the rods have a number from 0 to 9 at the top, and increasing multiples of that number written, in a characteristic diagonal fashion, along the rod from top to bottom. When the rods are placed next to each other in a frame, the products of multiplication sums can be read off, although the user still has to add up the digits for each place value. Napier explained his method of using the rods in his last work, *Rabdologiae*, published in 1617. He later developed a more elaborate version, a device he called the *promptuary*.

Napier's bones were akin to a portable set of multiplication tables.

31 Slide Rule

ANYONE WHO IS OLD ENOUGH TO REMEMBER USING A SLIDE RULE IS LIKELY TO DO SO WITH AFFECTION. The elegant simplicity of this device, which allows complex calculations to be done in seconds, has scarcely been surpassed by the pocket calculator. But a slide rule requires the user to remember values and follow the progress of calculations rather than just type in numbers.

A slide rule is often described as the first analog computer since in its most basic definition a computer is a device (or a person) that performs calculations—generally complex ones.

Just as log tables allow two numbers to be multiplied together by adding values read from tables, the slide rule places numbers on a log scale and multiplies them by adding their lengths along the scale. The first device of this kind, with logarithmic scales sliding against one another, was invented by English mathematician William Oughtred in the early 1620s. The slide rule soon became established as the most complex calculator of its day with many variations developed. Although now obsolete, it has been said that between 1700 and 1975 every single technological innovation was achieved with the help of a slide rule.

32 Complex Numbers

THE INTEGER NUMBERS SUBSCRIBE TO A CERTAIN SET OF RULES, NOT LEAST THOSE SET OUT IN EUCLID'S AXIOMS. One derived rule is that when two negative numbers are multiplied their product is always positive. Therefore all square numbers are positive by definition. So then the question naturally arose—what is the square root of a negative number?

The square root of 4 ($\sqrt{4}$) is 2 ($2^2 = 4$). However $\sqrt{4}$ could also be –2 since ($-2^2 = +4$). Just to be clear, 2 x –2 = –4, and this answer is not a square number, since it is the product of two different values. However, by the 16th century solutions to complex equations involved the square and cubic roots of negative numbers and mathematicians were forced to imagine how that could be.

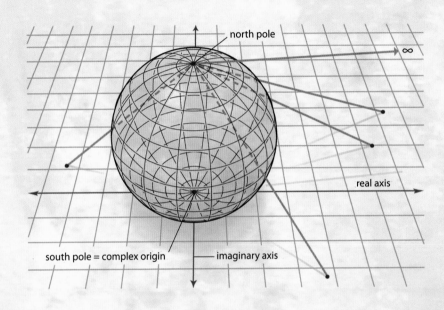

north pole

∞

real axis

south pole = complex origin ——— imaginary axis

In two parts

In 1545 Italian mathematician Girolano Cardano realized that although the square root of a negative number had no real number value, it could have an imaginary one, or as Cardano called it, a "fictitious" number. Soon after, Rafael Bombelli set out how the answers to some equations could be expressed in terms of what became known as complex numbers: in a complex number, there is a "real" component based on the unit 1 (which is also the $\sqrt{1}$) and an imaginary one based on the unit i. (We have René Descartes to thank for the term imaginary and Leonhard Euler for the use of the i).

The imaginary unit i operates just like a 1 ($i + 2i = 3i$) but forms the basis of a different set of numbers, which are entirely separate from real numbers. (There is no crossover between the two sets but otherwise the two are identical.) So a complex number can be ($1 + i$) or ($3 + 2i$). When adding or subtracting complex numbers, the real and imaginary parts are calculated separately. In multiplication the coefficient is applied to both sides. Carl Friedrich Gauss introduced the term complex number in 1831 described them as "a shadow of shadows," since he believed them to be just the simplest of a hierarchy of other imaginary quantities. In 1843 William Rowan Hamilton showed that complex numbers were a subset of quaternions, which add a fourth dimension.

In 1806, Jean-Robert Argand came up with the complex plane, where the real and imaginary components are plotted on perpendicular axes. This plot shows a three-dimensional Riemann sphere, one which contains all the complex numbers less than infinity (here shown as the north pole).

$$i^2 = -1$$

33 Cartesian Coordinates

RENÉ DESCARTES, (CARTESIUS TO HIS LATIN-SPEAKING AUDIENCE) IS WELL REMEMBERED FOR HIS HIGHLY QUOTABLE "I THINK, THEREFORE I AM", but he is also responsible for the coordinate system, which lets you know where you are once you have realized you exist. In actual fact, Descartes' mathematical contribution was not initially intended for use in maps and navigation, but as a means to unite geometry and algebra.

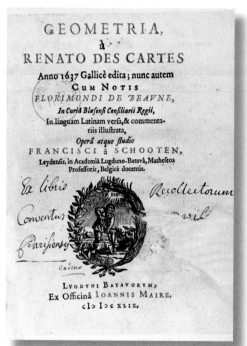

A Latin edition of Descartes' La Géométrie which was published as an appendix to his 1637 Discours de la Méthode. This work had a profound impact, leading almost directly to the development of calculus by Isaac Newton.

In true nerd style, René Descartes was a bit of a weakling, and so spent much of his youth in bed. His teachers allowed him to stay there until lunchtime, but his lie-ins were not idled away: far from it, he was the most able pupil at his Jesuit school in Paris. Perhaps unsurprisingly the habit of lounging in bed was a hard one to kick and Descartes continued it into adulthood. Much of his mathematical work set out in *Discours de la Méthode* (Discourse on the Method) in 1637 first occurred to him while serving in the Dutch army 20 years before (although still in bed).

Fly on the wall

The story goes that while in thought one morning, Descartes' eyes fell upon a fly crawling around on his wall and ceiling. He realized that while the route of the fly—and the shapes it figured—could be represented geometrically by tracing its path continuously, it could also be described as a series of points expressed algebraically.

He devised what is now called the Cartesian plane using two perpendicular number lines, or axes, to describe points on a flat surface. Descartes used a and b, but we now name the horizontal axis as x and the vertical as y. (Pierre de Fermat, frequently quietly critical of Descartes, independently formulated a three-dimensional system, with three axes. In case you were wondering, nowadays the third axis, such as the one used to plot a complex plane, is the *z*.)

A map reference is a familiar use of Cartesian coordinates, but they can also turn algebraic terms into lines, and turn shapes into algebra. A simple example is the formula of a straight line $y = mx + c$. m is the slope of the line, how much you multiply x by to get y, while c is the point where y equals 0, and the line crosses that axis.

Cartesian coordinates are always written in the form (x,y). The crossing point of the axes is called the origin, and has the coordinates (0,0).

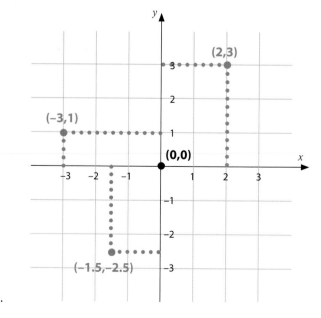

34 Laws of Fall

THE WORK OF GALILEO GALILEI MARKS THE BEGINNING OF A NEW TRADITION: THE FULL-ON APPLICATION OF MATHEMATICS TO SCIENCE. This broke with the existing tradition, grounded in the thinking of Aristotle, of debating qualitative explanations for natural processes—including falling—and made Galileo one of the first thinkers to believe that the laws of nature are written in the language of mathematics.

Galileo used experiments and measurements to describe motion with a precise mathematical law. This approach was perhaps more revolutionary than any of the quantifiable discoveries that Galileo is credited with. Some earlier investigators had already proposed the "law of squares"—the distance a body falls under gravity is proportional to the square of the time it takes to fall. It was also becoming understood that bodies continue in motion even when there is no force pushing or pulling on them.

Galileo's approach was to reject the notion of impetus as the reason why bodies continue in motion, along with Aristotle's idea that heavier bodies fall to earth faster because they have more of the element earth in them.

$$x = at^2$$

Galileo's law of fall states that the distance any object of any mass falls is proportional to the square of the duration of its fall: a ball falling for two seconds falls four times further than the same object falling for one second. The a in the law is a constant factor: later shown to be the acceleration due to gravity.

Instead, Galileo grounded his discoveries in painstaking experiment, and formulated his conclusions as mathematical laws. As well as firmly establishing the law of squares for falling bodies, he showed that their speed was directly proportional to the duration of falling. He also deduced that, for a projectile thrown upwards at an angle, the curve of its motion is a parabola (a conic section) resulting from a combination of its constant horizontal motion with its changing vertical motion.

The most complete account of Galileo's mathematical laws of physics is his 1638 book, *Discourses and Mathematical Demonstrations Concerning Two New Sciences*. By this time Galileo was living under house arrest, having been condemned in 1633 for promoting the heretical idea that the Earth moved around the Sun, and so this masterwork had to be smuggled out to a publisher in the Netherlands.

GALILEO AND THE LEANING TOWER

One of the most enduring myths about Galileo is that he demonstrated the facts behind falling by dropping spheres of greatly different weights from the Leaning Tower of Pisa. When the lighter sphere reached the ground at the same time as the heavier one, this at a stroke disproved the prevailing theory from Aristotle, that the rate of fall is greater for heavier objects.

This is almost certainly a thought experiment, rather than one actually performed by Galileo. It is mentioned not in Galileo's own writing but only in a biography by his pupil Vincenzo Viviani. It is likely that the experiment had been performed at some point before Galileo's time anyway, showing that it took more than one piece of contradictory evidence to break away from the Aristotelian world view.

35 Calculators

THE COMPUTER AGE HAS MANY STARTING POINTS, BUT PERHAPS THE FIRST MILESTONE was the Pascaline, a mechanical calculator that automated math—users no longer needed to understand the process to get the right answer.

The word *computer* hails from the year 1613 as a reference to a person whose job it was to perform complex calculations. Blaise Pascal's father was something of a computer, tasked with reorganizing the tax system of their French province. The young Blaise saw the advantage of having a automatic calculator and in 1642 began to develop what became the Pascaline. He built 50 versions before perfecting the device in 1645, then built approximately 20 machines of which nine survive. Society never looked back—today everything from a cash register to a spreadsheet does the math for us.

Numbers of up to six digits were inputted to the Pascaline by turning the dials. Inputting a second number results in it being added to the first one.

User interface

The Pascaline could be set for addition and subtraction, while multiplication was achieved by repeated additions. Numbers were dialed in by turning the wheel a fixed amount. The number was revealed in the windows above the wheel. Dialing in a new number resulted in it being added. If the result required it a figure was carried to the next column by a system of gears and escapements. Moving a bar made the machine subtract the numbers. Not every machine was in base ten, some counted in several bases for adding non-decimal currency or archaic distance measurements.

TIMELINE OF CALCULATORS

Abacus	**Slide rule**	**Stepped reckoner**	**Scheutz Difference Engine**	**Anita Mk 8**	**Hewlett Packard HP65**
Combination of counting rods and counting table	Perfected by William Oughtred	Invented by Gottfried Leibniz	First printing calculator	First commercially available electronic calculator	First programmable, handheld calculator
*16*th century	*1633*	*1673*	*1853*	*1961*	*1974*
1600	*1642*	*1822*	*1874*		*1971*
Napier's bones	**Pascaline**	**Difference Engine**	**Odhner pin wheel calculator**		**Busicom 141-PF**
John Napier's calculating kit	The first mechanical calculator	Never built but considered first computer.	The system used by all later mechanical calculators		First calculator with microchip

36 Pascal's Triangle

ALTHOUGH IT APPEARS TO BE A CHILDLIKE ARRAY OF NUMBERS, BLAISE PASCAL (AND MANY OTHERS) put this triangle together to explore the relationships between binomial coefficients. These are integers that combine in set theory, and their triangle is like a math kaleidoscope that provides glimpses into the heart of numbers.

```
                    1
                   1  1
                  1  2  1
                 1  3  3  1
                1  4  6  4  1
               1  5 10 10  5  1
              1  6 15 20 15  6  1
             1  7 21 35 35 21  7  1
            1  8 28 56 70 56 28  8  1
           1  9 36 84 126 126 84 36  9  1
          1 10 45 120 210 252 210 120 45 10  1
         1 11 55 165 330 462 462 330 165 55 11  1
        1 12 66 220 495 792 924 792 495 220 66 12  1
       1 13 78 286 715 1287 1716 1716 1287 715 286 78 13  1
```

This triangular figure was not invented by Blaise Pascal, but his work on it in the 1650s has led us to name it for the Frenchman. In fact the triangle was probably first assembled by Jia Xian in 11th-century China. Putting binomial theorem aside, it is simple enough to construct the triangle. The sum of two neighboring numbers results in the number immediately below both of them. The 1s that form the outer edge do not have outside neighbors (0 + 1 = 1). The second diagonal forms the counting numbers, since each is the sum of its predecessor and 1.

Pascal's Triangle has an outer skin of 1s and an underlay of natural (counting) numbers. It can also be imagined as being surrounded by an infinite ocean of 0s.

Looking deeper

There are many other patterns to be found. The third diagonal onwards is where the fun really starts. Its members, 1, 3, 6, 10... are the triangular numbers, meaning they can be used to construct two-dimensional triangles. The fourth diagonal has tetrahedral numbers, which form the subunits of tetrahedra, in effect three-dimensional triangles. The fifth diagonal has the pentascope numbers, which form hyperpyramids (four-dimensional triangles) and so on, adding a spatial dimension each time. There is more: shifting the triangle so its diagonals form into columns and then adding its "new" diagonals results in the Fibonacci sequence!

The triangle has some surprising patterns within it. The sum of every row is double its predecessor. Each one also contains the powers of 11.

Patterns within the triangle

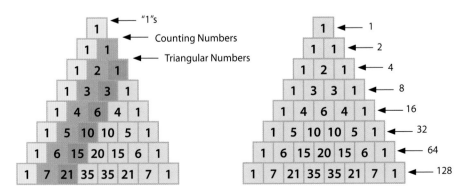

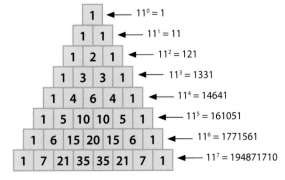

37 Chance

PEOPLE HAVE BEEN PLAYING GAMES OF CHANCE SINCE PREHISTORIC TIMES. However it wasn't until two great French mathematicians—who never actually met —exchanged a series of letters in 1654 that the mathematical study of chance, or probability, began.

Historians of mathematics have puzzled as to why understanding the laws of chance came so late on in mathematical history. Gambling by throwing dice or other objects (fancy a game of goat knuckle bones anyone?) goes back to ancient times. Apparently games of chance, such as drawing lots, were also used for divination—to find out "the will of the gods" and try to read the future. There may have been a feeling that it was somehow wrong to seek laws for such things, or perhaps people just thought that the future was impossible to predict by reason. From medieval times onwards there are occasional writings on the subject—a listing of the 36 possible results of throwing two dice, for example—and even Galileo wrote an unpublished article about a dice-throwing problem. But it was the French mathematicians Blaise Pascal and Pierre de Fermat who took thinking about chance to a new level.

One of the keys concepts of probability is to assume that each event—a roll of the dice, for example, occurs independently of those that occur before or after.

CHEVALIER DE MERE'S PROBLEM

Chevalier de Mere accepted bets that he would roll at least one six in four consecutive throws of a single die. He knew that the chance of rolling a six was 1/6 in a single roll.

Therefore he reasoned that his chance of rolling six in four tries was $(1/6) \times 4 = 2/3$, in other words more likely than not. However, the gambling Frenchman was not quite correct.

The total number of possible results from the four dice rolls is 6^4:

$6 \times 6 \times 6 \times 6 = 6^4 = 1,296$

The total number of possible losing results in which a six is not seen (just 1, 2, 3, 4, 5) in four rolls is 5^4:

$5 \times 5 \times 5 \times 5 = 5^4 = 625$

This figure lets us calculate the total number of winning results from four rolls:

$1296 - 625 = 671$.

Since 671 is greater than 625, the number of winning events is higher than the number of losing events. The Chevalier de Mere's betting position benefited only from a small advantage.

Exchange of letters

The initial stimulus for the famous series of letters between Pascal and Fermat came from problems posed to Pascal by a gambling friend, the Chevalier de Mere. The most significant of these is called "the problem of points." Two people are playing a game of dice, and the first to win a certain number of rounds wins overall, but in fact they decide to end the game early, with one player slightly ahead. How should they split the money fairly between the two participants?

The challenge of this problem is to think ahead accurately to all the possible outcomes that might have happened if the game had been completed. As with many probability problems, it is easy to jump to the wrong conclusion. The letters show Pascal struggling to solve the problem, and in fact it was Fermat's clearer mind that identified the correct solution.

The letters were quickly circulated around the members of the intellectual salon in Paris that Pascal belonged to. People could see at once that here was a new branch of mathematics, just waiting to be developed. What is more, although some problems could be solved with just a clear head and a counting up of possibilities, others were far too complicated and would require the creation of new mathematical methods to deal with them.

Degree of beliefs

The first of many treatises on the subject was written almost immediately by the Dutch scientist and mathematician Christiaan Huygens, who acknowledged the Pascal–Fermat letters as his inspiration. Then came Jacob Bernoulli, eldest of the famous family of Swiss mathematicians, whose posthumously published *Art of Conjecture* (1713) was the first to use the word *probability* in its modern sense. (Previously the word had just been used to mean "degree of belief.") Shortly afterwards, the Frenchman Abraham de Moivre showed how natural phenomena frequently averaged out into a bell-shaped curve, later named the normal distribution by Carl Gauss (the oft-styled Prince of Mathematics).

Much of this early probability theory was pure mathematics, or at most referred to games of chance that were already highly studied. But there were soon attempts to apply probability laws to the complexities of the real world. An early use involved predicting life expectancies for insurance companies, as reliable data was becoming available. It was not, however, until the late 19th century that probability theory was fully applied to analyzing statistics, a use that has continued to the present.

The chances are...

In the 20th century the mathematical theory of probability was recast in a more rigorous form. Also, a completely different approach to probability, the Bayesian method, finally became a practical possibility with the advent of computers, and is now widely used in risk management and decision making.

Probability theory, however, also remains just as relevant to its original context of games and gambling. Governments use it to help check that gambling machines are fair, for example. It is often the basis for mathematical puzzles, where it can still throw up considerable surprises.

Performing thousands of coin tosses is the only way to tell if this is a truly random process that is a fair method of making decisions.

THE MONTY HALL PROBLEM

This problem takes its name from the host of a TV game show. In the show, contestants had to guess which of three closed doors concealed a valuable prize. (The other two doors hid booby prizes.) The contestant chooses one of the doors, which remains shut. The host opens one of the two other doors to reveal a booby prize, and gives the contestant the choice of sticking to their original choice of door, or switching to the other unopened one. Common sense says the choice doesn't matter—it's now just 50/50. But amazingly, probability theory proves that the contestant should always switch their choice at this point to increase their chances.

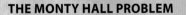
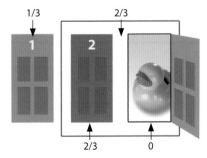

The selected door 1 has 1/3 chance of being correct. The two unpicked doors (door 2 and 3) have a combined probability of 2/3.

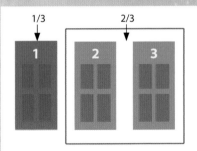

Opening door 3 (revealing an unwanted prize) does not reduce the combined probability, and so door 2 has a 2/3 chance of being correct. Always change your mind.

38 Principle of Induction

DEDUCTION USES GENERAL RULES TO DRAW CONCLUSIONS ABOUT SPECIFICS, **WHILE INDUCTION** uses particular instances to reveal generalities. In mathematics, induction is a form of proof, as rigorous as any other.

Although the term *induction* was coined in 1655 by the English mathematician John Wallis, the idea of inductive proof goes back to ancient Greece. In 1654, French mathematician Blaise Pascal provided a method of mathematical induction to prove the properties of the structure now known as Pascal's triangle.

$$0 + 1 + 2 + \cdots + n = \frac{n(n + 1)}{2}$$

Induction proves that this statement, known as P(n), shows that the sum of all integers up to n will equal the right-hand formula.

The idea of proof by induction is to infer that, if a mathematical statement or equation is true for a given whole number, this can be extended to all greater whole numbers. The basic strategy is as follows: show that the statement is true for some initial integers (usually 0, 1 or 2). Next, form the induction hypothesis—that the statement is true for any integer (call it n). Then, show algebraically that, if the statement is true for n, then it must also be true for the next number up ($n + 1$). This means, because the statement is true for the initial integer, it must also be true for the next one—and the next all the way to infinity.

39 Calculus

CALCULUS HAS BEEN CALLED "**T**HE MOST EFFECTIVE INSTRUMENT FOR SCIENTIFIC INVESTIGATION THAT MATHEMATICS HAS EVER PRODUCED." The problems that calculus addresses had been pondered by mathematicians since ancient Greece, but only piecemeal answers had been found. Then within a few years of each other, two great minds separately proposed solutions.

Newton's approach to calculus was to use tangents— the exact slope of a curve at a given point.

The founders of calculus were Isaac Newton in England, and Gottfried Wilhelm Leibniz in Germany. While their solutions were equally powerful mathematically, their approaches were distinctly different.

f(x) — tangent — arbitrary point *x,y* — *y* — *x*

Hidden discovery

Newton was primarily interested in measuring rates of change of speed at instants in time, or infinitesimals. This was in order to understand the motions of the planets, which change speed throughout their orbits. Newton was a superb mathematician who had already made important advances in summing infinite series before tackling this problem. In

developing his method of calculus, which he named *fluxions*. Newton was aware of the logical problems associated with infinitesimal quantities—how could they exist at all if they were infinitely small? This unease may be one reason why, although he seems to have developed calculus late in 1665, he did not publish his discovery then. Even some 20 years later, in his great work known as the *Principia*, Newton avoids setting out his complete method of fluxions and refers instead to "first and last ratios"—an idea similar to the later mathematical concept of *limit*. Newton was perhaps right to be cautious, as a logically rigorous foundation for infinitesimal calculus was not achieved until the mid-19th century.

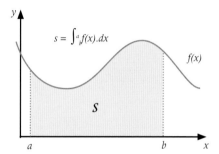

Leibniz's approach involved integration. This process was applied to a function to determine areas and volumes. Here, the function is the red line f(x). The area S under the curve is the integral of f(x) between the limits 'a' and 'b', which is the integrated function of x=b, minus the integrated function of x=a.

$$s = \int_a^b f(x).dx$$

A table published in 1779 compares various simple curves from calculus with elliptical and hyperbolic curves from the analysis of conic sections.

Loud and clear

In contrast, Leibniz was a diplomat and a philosopher—parodied in Voltaire's *Candide* as the absurdly optimistic Dr Pangloss—who only became interested in mathematics later, around 1672. For Leibniz, infinitesimals were real quantities, so he had none of Newtons's qualms about publishing his method of calculus in 1684. Leibniz was initially interested not in rates of change but in how the area of shapes bounded by curves can be precisely calculated. Ingenious methods of approximation, based on packing the area with smaller and smaller shapes with calculable areas had been used since the days of Archimedes, but Leibniz's method provided the long-sought general solution. He also introduced the notations used in calculus: *dy/dx* for rates of change (differentiation), and ∫ (a long s) for summing areas (integration).

Common ground

Despite their differences, both Newton and Leibniz arrived at the fundamental theorem of calculus: that the mathematics of integration and rates of change are so closely related that each is in fact the inverse of the other. For example, imagine a car accelerating from zero to its top speed. The rate of change of position is, of course, the speed; and the rate of change of speed is the acceleration. Conversely, integrating all the speeds—and the time spent traveling at them—will give the total change in position (distance traveled), while integrating all the rates of acceleration will give the final speed. It is this power to pin down mathematically the relationship between different changing quantities that gives calculus its vast scope.

THE PRIORITY DEBATE

Leibniz and Newton became locked in perhaps the most vitriolic argument in the history of math: who invented calculus? Although Newton did not publish his ideas in full until 1704 , Newton's notes show that he had developed them decades earlier than his rival. The dispute produced a rift between English and European scientists. Leibniz's appeal to the Royal Society of London (below) had no effect—Isaac Newton was its president. Although Newton got there first, we use Leibniz's calculus.

40 Math of Gravity

MOST PEOPLE WILL HAVE HEARD THE STORY OF HOW ISAAC NEWTON WAS DISTURBED BY AN APPLE FALLING IN HIS GARDEN AND WONDERED. "Why doesn't the Moon fall to Earth like the apple?" Whether or not this actually happened, it was Newton's genius to realize that the force that holds the moon in orbit is the same force that pulls the apple to the ground.

Everything in the Universe produces a gravitational force that pulls other objects towards it and it is the force of gravity that determines the paths that the stars, planets, and other objects follow through space. Newton had studied Galileo's work on projectiles, and he suggested that the Moon, or indeed any other object in orbit, could be considered as a projectile. A projectile follows a curved path as it is pulled back to earth by gravity. The surface of the Earth is also curved, so it follows that, if a projectile is traveling fast enough, its curved path will follow the curve of the Earth. The projectile is still falling, but now it is falling around the Earth and has become a satellite. If the velocity of the projectile is increased still further the path it takes around the Earth becomes an ellipse. Therefore in effect, a planet in orbit is falling around the sun.

$$F = G\,\frac{m_1 m_2}{r^2}$$

Newton and his apple is one of the most famous stories in science. The great man kept quiet about it until very late in life, when he had mellowed somewhat from his reclusive, acerbic, and downright rude persona of his more active days.

The force of gravity (F) between two masses is calculated as the product of both masses (m), divided by the square of the distance between them (r) multiplied by the gravitational constant (G).

Inverse square

Newton knew that the force of gravity causes falling objects near the Earth's surface (such as the famous apple) to accelerate towards the Earth at a rate of 9.8 m/s². He also knew that the Moon accelerated towards the Earth at a rate of 0.00272 m/s². If it was the same force that was acting in both instances, Newton had to come up with a plausible explanation for the fact that the acceleration of the Moon was so much less than that of the apple. What characteristic of the force of gravity caused the more distant Moon's rate of acceleration to be a mere 1/3600th of the acceleration of the apple?

It seemed obvious that the force of gravity was weakened by distance. But what was the formula for determining it? An object near the Earth's surface is approximately 60 times closer to the center of the Earth than the Moon is. (It is roughly 6,350 km from the surface to the center of the Earth and the Moon orbits at a distance of 384,000 km from the Earth.) The Moon experiences a force of gravity that is 1/3600 or 1/(60)² that of

the apple. Newton realized that the force of gravity follows an inverse square law (6350 x 60 ≈ 384,000)

The force of gravity between two objects is inversely proportional to the square of the distance that separates those objects, whether they are the Earth and an apple, the Earth and the Moon, or the Sun and Mars. Double the distance and the force is reduced to a fourth, triple the distance and the force reduces to a ninth, and so on. The force is also determined by the masses of the objects—the greater the mass the greater the force of gravity. In summary, the force of gravity is proportional to the product of the masses of the objects and inversely proportional to the square of the distance between them. (The constant of proportionality is the conveniently titled gravitational constant)

Understanding gravity allows us to do extraordinary things, like measuring the mass of the Earth. In 1798, by careful experiment, Henry Cavendish succeeded in making an accurate determination of G, the gravitational constant, as 6.67×10^{-11}. According to Newton's theory of universal gravitational attraction, G determines the gravitational attraction between any two masses anywhere in the universe. This meant that the mass of the Earth could now be determined. A 1 kg mass at the Earth's surface is approximately 6,300,000 meters from the center of the Earth and the force acting on it is approximately 10 Newtons. So, by inputting these figures into the gravity equation we can find that the mass of the Earth is roughly 6×10^{24} kg.

THE SHOULDERS OF GIANTS

Acknowledging his debt to other scientists, Newton famously remarked, "If I have seen farther it is by standing on the shoulders of giants." This is regarded as a nod to the work of Galileo and Kepler. Galileo determined that the motion of a projectile had two components—the uniform acceleration acting vertically on the projectile and the horizontal motion, which is motion at a steady speed in a straight line. Newton pointed out that the vertical acceleration of the projectile must be the result of a force (gravity) acting on it; otherwise it would fall at a constant speed. However, Newton's statement is also thought to be a jibe at Robert Hooke (right), who demanded credit from Newton for some ideas—and was very short in stature.

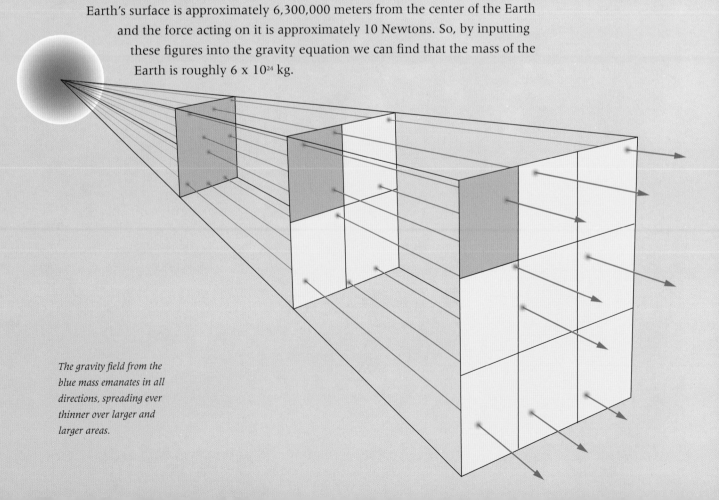

The gravity field from the blue mass emanates in all directions, spreading ever thinner over larger and larger areas.

41 Binary Numbers

THE WORD *DIGITAL* IS EVERYWHERE TODAY, PERHAPS OVERUSED. IT HAS LONG IMPLIED A CERTAIN CACHET from the one-time glamor of a digital watch to a crystal-clear digital radio signal. However, digital really just means "using numbers" and more often then not just two numbers, 0 and 1. Welcome to the binary system.

The representation of the binary system set out by Leibniz in this 1703 publication of his paper Explication de l'Arithmétique Binaire *is basically the one used today.*

Binary is a positional counting system that uses just two numerals instead of the ten we are more familiar with. What could be more simple? Well, it may be easier to learn then your 1,2,3s—just stop at one and you are done—but binary numbering soon becomes rather unwieldy in real-world applications: counting just the fingers on one hand produces a binary number of 110. The number of players on a soccer field (22 for the uninitiated) is written as 10110. Add the three referees and you get 11001. Clearly counting in binary is far from intuitive. So why bother?

Seven years after he rocked the math world with his version of calculus in 1672, Gottfried Liebniz's contribution to the use of the binary numeral system was less controversial but no less profound.

Two-state solutions

The benefit of binary is most apparent in the spooky world of codes and cyberspace. This was spotted by a remarkably prescient Francis Bacon in 1605. In between his day job of being a corrupt lawyer and courtier to the English monarch, Bacon achieved a lot, not least sparking the 17th-century's Scientific Revolution with his scientific method. He also saw the potential of binary numbers when he explained that the entire alphabet could be ciphered using strings of five binary characters. (There are 26 letters and the total permutations of five characters is 32 (2^5).

Bacon's genius flashed into action when he recounted how these codes were not limited to the written word, but could be transmitted by any method with a "twofold difference onely; as by Bells, by Trumpets, by Lights and Torches, by the report of Muskets, and any instruments of like nature." Samuel Morse's telegraph code of dots and dashes certainly owes a debt to this notion, and in the modern world, the switch-like transistors of a microprocessor are just a such a *twofold instrument*.

Ones and zeros

Bacon's cipher used *a*s and *b*s rather than numerals—"A" was coded as *aaaaa*—but the principle was the same nonetheless. It is another science superstar we have to

HEXAGRAMS OF THE *I CHING*

Gottfried Leibniz was an avid orientalist, endlessly fascinated with the mysterious east. Just like many similarly minded people after him would, his research took in the *I Ching*, one of the oldest works of Chinese literature, dating from 1,000 BC if not before. It is concerned with divination, predicting the future with a series of symbols called trigram and hexagrams. The eight trigrams—the ones often seen arranged around the *yinyang* symbol—are made of three lines, while the 64 hexagrams (below) have six lines. An unbroken line represents *yang*, a broken one is *yin*, opposite characters of an interconnected whole: Leibniz saw them as something more—binary 0s and 1s. That gave the hexagrams a numerical value up to 2^6 (64). The trigrams counted up to 2^3 (8). Hexagrams are in effect doubled trigrams ($2^3 \times 2^3 = 2^6$).

-0-1-2-3-4-5-6-7-8-9

thank for the current binary notation. Gottfried Leibniz, one of the founders of calculus, introduced the 0 and 1 digits in his *Explanation of Binary Arithmetic*, published in 1679.

While it requires a good deal of practise to read binary numbers, they are still set out in the same way as described by Leibniz, no different in fact to any decimal number.

Read right to left, the first figure of a decimal number, let's say 31, is the units (1), and the next is the number of tens (3). Decimal numbers continue into the hundreds, thousands, and so on. Keen observers will know that each additional number position is 10 to the next power: the units are multiples of 10^0, which equals 1, the tens are 10^1, hundreds are 10^2, and thousands 10^3. The binary system merely replaces 10 with 2. The number begins with the units: 2^0 or multiples of 1. The next position is 2^1, which is simply 2 in decimal but 10 in binary. The positions that follow are 2^2, 2^3, 2^4, which are 4s, 8s, and 16s. So the decimal number 31 is the binary number 11111.

One additional notation has been introduced since Leibniz, relating to any number system with a radix different from 10. To explain some terminology: the radix is the number of digits used by the system (counting the zero). This is also frequently termed base, so a decimal system has a radix of ten and its numbers are in "base ten." Binary numbers are in base two and to make this clear the base (or radix) is shown in subscript after the number. Therefore 11111_2 = 31_{10}. (After 10 and 2 the most common radix in use is 16. This hexadecimal system has digits 0–9 followed by A–F; $F_{16} = 15_{10}$, $DEC0DE_{16} = 14,598,366_{10}$.)

Converting binary

To convert a binary number into a decimal one, every numeral is replaced with the power of two relating to that position and then added together. The digit on the right is equal to 2^0, the next one is 2^1, followed by 2^2 and so on. For example, the binary number 1010101_2 would become: $1 \times 2^0 + 0 \times 2^1 + 1 \times 2^2 + 0 \times 2^3 + 1 \times 2^4 + 0 \times 2^5 + 1 \times 2^6 = 1 + 0 + 4 + 0 + 16 + 0 + 64 = 85_{10}$.

To write a decimal number in binary, the number should be divided by two repeatedly until the answer is 0. For example, 50_{10} is 110010_2: 50 / 2 gives 25 (remainder 0); 25/ 2 gives 12 (remainder 1); 12 / 2 gives 6 (remainder 0); 6 / 2 gives 3 (remainder 0); 3 / 2 gives 1 (remainder 1); 1 / 2 gives 0 (remainder 1) and we stop. This first remainder is the number of units (2^0) and the other remainders follow on: $0 \times 2^0 + 1 \times 2^1 + 0 \times 2^2 + 0 \times 2^3 + 1 \times 2^4 +$ that final remainder 1×2^5. Rearranged this gives us the number 110010_2.

Decimal numbers	Binary numbers
0	0
1	1
2	10
3	11
4	100
5	101
6	110
7	111
8	1000
9	1001
10	1010
11	1011
12	1100
13	1101
14	1110
15	1111
16	10000
17	10001
18	10010
19	10011
20	10100

NEW NUMBERS, NEW THEORIES

42 *e*

THE MATHEMATICAL CONSTANT *e* IS A RELATIVE NEWCOMER—a brash interloper on a field previously dominated by ancient numbers, such as pi and phi (the golden ratio). However, this fascinating number now holds a position at the very heart of mathematics.

e stands for *exponential growth constant*, and it is nothing more than a number. Even so, no other number has more diverse definitions. The number is irrational (and indeed transcendental)—a trailing tail of decimals that never ends.

Unlikely though it may seem, this number applies to countless real-world situations. It is most often allied to growth functions, phenomena such as radioactive decay, accumulation of capital, epidemiology and the flourishing of bacterial colonies. Euler's number also has an uncanny knack of popping up in the fundamental relationships of mathematics. Why such a number should occur in so many surprising places is one of *e*'s many mysteries and delights.

Although it appeared in a table in John Napier's book, it was thought the first use of e in calculating natural logarithms was the work of English mathematician William Oughtred.

Natural-born rhythm

Contrasted with π's long association with geometry, *e* traces the birth of modern mathematics. The first glimpse of the *e* came in 1618, tucked away in an appendix to John Napier's work on natural logarithms. Napier's advances with logarithms made calculating large and complicated multiplications a doddle. They provided a model for changing multiplication into addition [$\log_n(xy) = \log_n(x) + \log_n(y)$] and were eagerly adopted by mathematicians. *e* arises from the construction of the so-called natural logarithm, which defines a new number—a fixed-value "base" the natural logarithm of which is 1. The natural logarithm of any given number x is then the power to which this base must be raised to equal x. In other words, $\ln(e^x) = x$ and $\ln(e) = 1$, since $e^1 = e$.

This logarithmic function is natural in mathematical terms since it is defined around oneness. As you can see, *e* is the public face of the exponential function, e^x, where x = 1. The natural logarithm and the exponential function are twins—two sides of the same coin—e^x is the inverse of the $\ln(x)$, and *e* is the inverse of the $\ln(1)$. The

e to 50 decimal places. There is a lot more to it than you see here.

2.71828182845904523536028747

exponential function can also be defined as an infinite power series: $\sum [\infty] [n=0]\, x^n/n! = x^0/0! + x^1/1! + x^2/2! + x^3/3! + x^4/4! + \ldots = e^x$. ($\sum$ means "the sum of"; ! means "factorial," where a number is multiplied by every other integer less than it). This series has the unique property that on differentiation none of the terms disappear, and instead repeat ad infinitum. So the exponential function is its own derivative and the graph of e^x describes its own rate of change. This explains why this number is so ubiquitous in mathematics and how it tends to crop up everywhere in calculus.

Defining e

The Swiss Jacob Bernoulli was the first to reveal the value of e by exploring the nature of compound interest in the late 17th century. A simple 100% increase when paid annually results in the sum doubling year on year. But to split the 100% over shorter periods results in more than double the return at the end of the year (>2). Bernoulli calculated what the payout would be if the interest was calculated continuously over infinitely small periods (try telling that to your bank) and the answer was it grew at the annual rate of 2.71828… or e. In this guise the amazing e is found in other situations that involve growth or decay, such as radioactivity, bacterial infections, and epidemics.

The members of the mathematical Bernoulli family were known as the Kings of Basel during the 17th and 18th centuries. Their work permeates through mathematics from fluid dynamics and calculus to probability. Here, Jacob explains the constant e to his brother Johann.

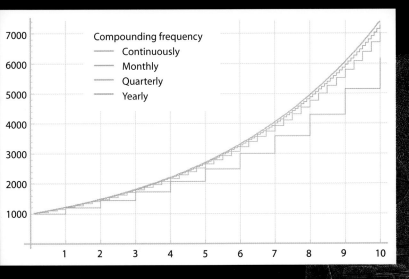

The rate of growth of savings is proportional to the frequency of compounding.

135266249775572470936 9995…

43 Graph Theory

THIS AREA OF MATHEMATICS APPEARED ALMOST OUT OF NOWHERE IN THE 18TH CENTURY. HOWEVER IT SOON FORMED A BRIDGE BETWEEN GEOMETRY AND more esoteric fields of enquiry, such as topology, combinatorics, and set theory, using the properties of objects to reveal deeper truths.

An engraving of Königsberg from the 16th century only shows six of the seven bridges, the last one was to be built behind the couple in the picture.

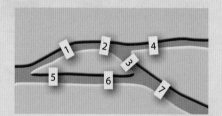

It will come as no surprise to find Leonhard Euler at the heart of graph theory. In fact he was there at the very beginning, taking a vacation in the Baltic city of Königsberg. The theory took root in Euler's classic paper of 1736 entitled *The Seven Bridges of Königsberg*.

Making connections

A popular diversion for the residents of Königsberg (then a Prussian city but now named Kaliningrad and located in the Russian enclave between Poland and Lithuania) was to stroll around town attempting to cross each of the city's seven bridges just once. Euler may have tried himself, or perhaps realized the mathematical futility. No one had ever managed to find a route between the two islands and the mainland, and many thought it impossible. But if it was truly impossible, surely there was a reason—there must be a proof. As Euler put it, the problem "seemed worthy of attention because neither geometry or algebra or the art of counting was sufficient to solve it."

Graphic language

Euler's insight was that the key is the number of bridges connecting each area; that magnitude and measurement are unnecessary. In the context of this theory a graph is not a line on a Cartesian plane; coordinates are irrelevant. Instead a graph is a collection of points, or vertices, that are connected by lines, or edges. If the edges are given a length, the graph is said to be weighted, and if a direction is attributed to an edge it becomes known as an arc.

Three very different shapes but three identical graphs. Count the vertices (dots) and edges (lines).

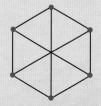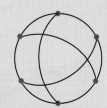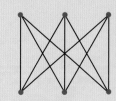

Metro maps are just one real-world example of a graph. For a traveler it only matters how the stations are connected and not how far they are away from each other.

The Königsberg problem has four edges, the regions of the city connected by bridges, and seven vertices, the bridges themselves. Euler showed that there is no route covering each edge once because the graph has more than two vertices with an odd number of edges.(The number of edges on a vertex is said to be its degree.)

Therefore the bridge problem has no solution and history has removed the need for one since Kaliningrad lost several bridges in the Second World War. There are now just three bridges and an anonymous motorway overpass—the Leninsky Prospekt—runs right across the island once at the center of the problem.

The Bridges of Königsberg is an introduction to a theory which is now the sort of math behind computer programs that can recognise fingerprints and faces. It can also be used to design manufacturing processes and is still used to find optimal routes or plan physical networks (not least the Internet). Graph theory is also used to analyze the moves in games and is one of the reasons why a computer will normally beat you at chess.

44 Three-Body Problem

NEWTON'S LAW OF GRAVITY WORKS VERY WELL IN DETERMINING WHAT HAPPENS between two objects but adding just one more body to the equation results in incredible complexity.

Trying to compute the mutual gravitational attraction of three bodies is surprisingly difficult. In fact, the three-body problem is essentially unsolvable. In 1747 an approximate solution was found by treating the Sun as if it were a fixed object, allowing calculations to be made for the motion of the Earth and Moon as they orbit the sun together. Later Joseph-Louis Lagrange discovered that there were five special points in this Earth-Moon reference frame where the gravitational forces effectively canceled each other out. An object placed at any one of these Lagrange points would orbit the Sun but maintain the same relative position with respect to the Earth–Moon system.

The ongoing battle to solve the forces acting between the Sun, Moon, and Earth was one inspiration behind chaos theory.

45 Euler's Identity

WHEN READERS OF *MATHEMATICAL INTELLIGENCER* MAGAZINE were asked, in 2004, to vote for the "most beautiful theorem in mathematics, the winner—by a considerable margin—was Euler's identity:

$$e^{i\pi} + 1 = 0$$

Leonhard Euler was born in the German-speaking region of Switzerland but spent most of his working life in Saint Petersburg, Russia.

Euler's identity written as an equality.

Formulated by its namesake in 1747, in total the theorem uses the five most important constants of mathematics: zero (0), unity (or 1), the exponential number e, which describes growth and decay, ($e \approx 2.718...$), π, the ratio of a circle's circumference to its diameter ($\pi \approx 3.142 ...$), i, the basic imaginary number which, if it did exist, could be squared to make −1. (It is imaginary in the sense that it is not "real;" neither positive, negative, nor zero, since the square of any positive or negative number is always a positive number, and the square of zero is zero.)

The man and his identity

Beyond more academic circles Leonhard Euler is not as well known as math superstars like Newton, Leibniz, Gauss and even his occasional tutors the Bernoullis. However, the influence of Euler—known as the Magician—can be seen running through number theory, calculus, and graphics. He was a prolific assimilator of mathematical terms and inventor of symbols. The capital sigma Σ is used to denote "the sum of" thanks to Euler. He also introduced the use of e and i, and popularized writing pi as π. Euler's identity combines all of these with unbeatable simplicity.

The identity refers to the properties of complex numbers: a complex number is the combination of an ordinary—or real—number and an imaginary one. Mathematicians had been tinkering with these numbers for a couple of centuries and Euler's equation allowed them to be used in functional analysis.

A basic proof

Complex numbers can be represented as co-ordinates on a plane in which the vertical height is the size of the imaginary part of the number and the horizontal distances are the real part. To reach this position, one may consider the rotation of a line anticlockwise from a horizontal position, which travels through an angle of x. In that case the position of the point is given by $\cos(x) + i\sin(x)$. This can be written in terms of e: e^{ix}. If the angle x is a half-circle (that is, π radians), then it becomes $e^{i\pi} = -1$, or $e^{i\pi} + 1 = 0$.

The expression e^{ix} plotted on a complex plane with $x = \pi$, the angle of a half circle when measured in radians.

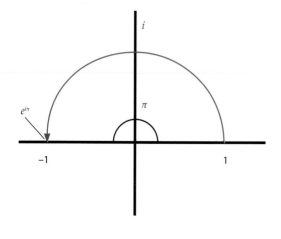

46 Bayes' Theorem

ON THE SURFACE, BAYES' THEOREM, DEVELOPED BY THE REVEREND THOMAS BAYES AND PUBLISHED POSTHUMOUSLY IN 1763, is an innocent enough formula—it simply calculates how the probability of an event changes in the light of new information being provided. However, what it shows about the world is both surprising and, in some cases, highly contentious.

$$P(A|B) = \frac{P(B|A)P(A)}{P(B)}$$

Bayes' theorem relates the probability of event A occurring, given that event B has taken place [P(A|B)], to the probabilities of A or B happening independently and the conditional chance of B happening given A.

Let's say that there is an outbreak of flu, and about one in 100 people have it. So, there is a one per cent chance that you have it. Then, you wake up with a headache and it says on a medical website that 90 per cent of flu sufferers have one too. So it sounds like you're unlucky and have caught the virus. But then again, you were out late at a party the night before and you think the probability that you would have a headache anyway is about 10 per cent—unlikely, but a definite possibility. So what is your chance of having flu now, given that you have a headache? Is it still 90 per cent or perhaps 80 per cent, taking the possible hangover into account? Bayes' theorem tells us it is just over 8 percent. Chances are you will be fine.

Continued debate

Bayes' theorem sparked off a debate in mathematics that continues to this day, with battle-lines drawn between the Bayesians and the frequentists. This is not to say that there is any disagreement about the validity of the theorem, but rather about its suitability in certain cases, where a person's beliefs are involved in setting prior probabilities.

In our flu epidemic examples, there would be no debate: the prior probabilities are the one per cent chance that you have the flu in the first place (i.e., prior to noticing your headache). However, establishing the probabilities of prior events are not so clear cut. If a doctor's first impression is that there is a 1 in 10 chance you have flu, then is the status of that 10 percent really the same as the observation that 1 in 100 people have the flu? A Bayesian would say yes, but a frequentist would decline to do so until a significantly large sample of similar situations had been recorded and compared.

UNDERSTANDING CHANCE

In a UK rape trial in the 1990s, the victim had failed to pick out the defendant in a identity parade and when asked, said he did not match the age or description of her attacker. Nevertheless, he was convicted of her rape on DNA evidence. The jury was told that only one in 20 million people could have this DNA. In a later retrial the defense used Bayes' theorem asking: "If he were the attacker, what's the chance that she would say he looked nothing like the attacker? And if he wasn't the attacker what's the chance that she would say he looked nothing like the attacker?" The idea was the jury would think the second condition was more likely. But the weight of statistics behind the DNA still swayed the jury. The suspect was reconvicted.

Identification by DNA is based on probability, an approach that is increasingly open to dispute, since juries are rarely given guidance on the likelihood that the DNA could also belong to a suspect's relatives.

47 Maskelyne and the Personal Equation

Nevil Maskelyne is also remembered for calculating the density of Earth by measuring the gravitational pull of a Scottish mountain. His result was 80 per cent of the true value.

WHEN NEVIL MASKELYNE, BRITAIN'S ASTRONOMER ROYAL, SACKED HIS ASSISTANT IN 1796 FOR APPARENTLY MEASURING STARS INACCURATELY, HE DID MORE THAN HE KNEW. Unwittingly, he initiated the important topic of how personal factors affect any kind of measurement.

By watching stars through a telescope and listening to a clock ticking, astronomers of Maskelyne's time aimed to measure events within a fraction of a second. Maskelyne thought his assistant was around half a second late most of the time, and published this opinion along with the mixed results of their observations. But, after Maskelyne's death, the German astronomer, Friedrich Bessel investigated. He discovered there was a regular, measurable difference between any two of his colleagues making observations, and this became known as the *personal equation*.

For astronomers it was mainly a practical problem, but later in the 19th century the equation stimulated detailed research on reaction times, within the new science of experimental psychology. Eventually the phrase caught on with the general public, and was applied—rather vaguely—to any personal factor in any situation.

48 Malthusianism

THOMAS MALTHUS'S NAME HAS BECOME INEXTRICABLY LINKED WITH HIS ARGUMENT THAT UNCHECKED POPULATION growth could only result in a catastrophic collapse as a result of famine, disease, and a desperate war of survival.

AN ESSAY
ON THE
PRINCIPLE OF POPULATION;
OR, A
VIEW OF ITS PAST AND PRESENT EFFECTS
ON
HUMAN HAPPINESS;
WITH
AN INQUIRY INTO OUR PROSPECTS RESPECTING
THE FUTURE REMOVAL OR MITIGATION OF
THE EVILS WHICH IT OCCASIONS.

By T. R. MALTHUS, A.M.
LATE FELLOW OF JESUS COLLEGE, CAMBRIDGE.

IN TWO VOLUMES.
VOL. I.
THE THIRD EDITION.

LONDON:
PRINTED FOR J. JOHNSON, IN ST. PAUL'S CHURCH-YARD,
BY T. BENSLEY, BOLT COURT, FLEET STREET.
1806.

Thomas Malthus's work was presented in this book, first published anonymously in 1798.

The Reverend Malthus set out this theory in 1798: since population grows geometrically it would eventually outstrip food production, which could only increase arithmetically, and the natural outcome would be a period of misery and death until the population fell to a sustainable level. A geometric sequence increases by multiplying by the same value 1, 2, 4, 8, 16… whereas an arithmetic sequence increases by adding the same value – 1, 2, 3, 4, 5…

Malthus argued that the recent Poor Laws passed in Britain, which provided an early form of welfare payments that depended on the number of children in a family, were wrong: first, it encouraged the poor to have more children because they would produce as many as they could feed. Swelling the numbers in the workforce would reduce labor costs, which would ultimately leave the poor worse off. Second, if money were being provided to every poor person by the government, manufacturers and services would raise their prices to take advantage of this.

Malthus's ideas were spurred by debates with his father about how "perfect" a society could be. He suggested that attempts to improve conditions for the poorer members of society were doomed to fail because better conditions would lead to increased population growth, which would eventually outstrip any increase in production. This, according to Malthus, put any idea of a "perfect" society beyond reach.

Future looks brighter

However, the future was not to be Malthusian, at least not yet. Malthus could not have predicted the changes that would come about as a result of the industrial revolution. Technological advances made food production more efficient with more food produced in smaller areas of land than could ever have been imagined in the 18th century. Malthus could also not have foreseen the impact of public health services and family planning using contraceptives.

Europe has a falling population because, counter to Malthus's ideas, prosperity appears to lead to lower birth rates. The world's current population growth rate is about 1.14 per cent. If sustained it would lead to a doubling of the population in about 61 years. However, the population growth rate varies—it peaked in the 1960s at 2.2 per cent, a rate that would have meant a doubling time of 35 years had it continued.

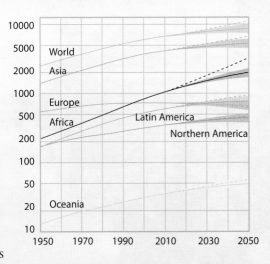

The logarithmic nature of the vertical axis (indicating billions) flattens the curves on this graph. Nevertheless the population of every region bar Europe is growing. Increased family planning is predicted to see the world population flatten out in the 2050s.

49 Fundamental Theorem of Algebra

Carl Gauss went on from his proof of the fundamental theorem to be a leading figure in statistics, probability, number theory, and astronomy.

THE FUNDAMENTAL THEOREM OF ALGEBRA SHOWS THE FIELD OF COMPLEX NUMBERS IS ALGEBRAICALLY CLOSED, or putting it another way there is no polynomial expression that does not have at least one root (x_n) which equals zero.

Factoring out $x–x_n$ leaves a polynomial of degree n–1. As Carl Gauss said: "Every polynomial equation of degree n with complex coefficients has n roots in the complex numbers." Lest we forget, complex numbers are those that have a real and imaginary component (based on the number i, the square root of –1). Real numbers, the ones we are all familiar with, are a subset of complex numbers (the imaginary part is $0i$).

More than a hunch

The first claim that there are always n solutions to equations of n degree with n roots was made by Flemish mathematician Albert Girard in 1629 but he left open the possibility that solutions could be found outside the complex numbers. In 1637 the philosopher René Descartes said that for every equation of degree n, n roots could be imagined but these imagined roots did not correspond to any real quantity.

The first serious attempt at proving the fundamental theorem was made by Jean d'Alembert in 1746. Though there were weaknesses in his argument—he used unproven statements that themselves depended on proving the fundamental theorem—his ideas were helpful.

GRUDGING RESPECT

Such is the rivalry among mathematicians that acknowledgement of Gauss's proof of the fundamental theorem was a long time coming, if at all. At the celebrations in Basel for the 1907 bicentenary of Leonhard Euler's birth Georg Frobenius said: "Euler gave the most algebraic of the proofs of the existence of the roots of an equation, the one which is based on the proposition that every real equation of odd degree has a real root. I regard it as unjust to ascribe this proof exclusively to Gauss, who merely added the finishing touches."

Princely proof

Credit for the first proof of the fundamental theorem usually goes to German Carl Friedrich Gauss. He presented his proof in 1799, while still a young man—barely 22. (He would later become the most influential mathematicians of his generation— the "Prince of Mathematics.") Gauss's work pointed out the fundamental flaws that had bedevilled earlier attempts at a proof but his own proof had gaps, and modern mathematicians do not consider it to be sufficiently rigorous. Interestingly, Gauss himself made no claims for his being the proper proof. He called it a "new" proof and recognized the value of d'Alembert's work.

In 1814 Swiss accountant Jean Robert Argand published one of the simplest of all proofs of the fundamental theorem, based on d'Alembert's ideas. Argand's proof was of a type known as an existence or nonconstructive proof, which indirectly shows that a mathematical object exists, but without providing a specific example of it. (It wasn't until 1940 that Hellmuth Knesser succeeded in producing a constructive variant of Argand's proof that could directly provide specific examples.) Two years later in 1816 Gauss published a complete proof building on the earlier work of Leonhard Euler in the 1740s. Gauss's second proof used indeterminates, symbols that do not stand for anything but themselves, whereas Euler had been operating with roots that may not exist.

50 Perturbation Theory

IN THE 17TH CENTURY, NEWTONIAN
PHYSICS held out the prospect that all the motions in the Universe would succumb to the power of mathematics, bolstered in particular by what we now call calculus.

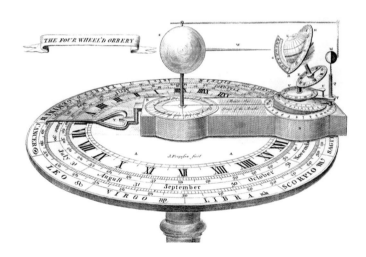

Mechanical models, or orreries, of planetary systems relied on the clockwork nature of the Newtonian Universe.

It rapidly became clear that although in principle the motions of the planets could be predicted exactly, the actual calculations were extremely complex. The three-body problem was a case in point: it proved impossible to find a computable solution to the gravitational interaction of Sun, Earth, and Moon.

Techniques of the perturbation theory were used to explain the observed motion of all heavenly bodies around Earth via a complex system of eccentric circles and epicycles.

Newton found a practical approach to such situations: the dominant influence on the Moon was clearly the Earth, so other effects could be regarded as perturbations to that influence. Hence, Newton changed the question from the intractable "what is the result of the interaction between Earth, Moon, and Sun," to the more calculable "what change does the Sun make to the Earth–Moon interaction?" (In effect this principle had been used before, in the addition of epicycles to circular orbits by Ptolemy and others to approximate the measured motion of the planets, but in that case the perturbations were thought to be real, not mere conveniences.)

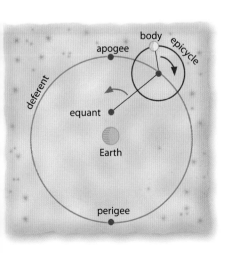

Perturbation theory has many other applications too. One of its greatest triumphs was published in 1799 by Pierre-Simon Laplace. By making small perturbing changes to the model solar system he had set out, Laplace found that it remained stable, confirming that the model was rugged enough to be a true reflection of reality.

51 Central Limit Theorem

THE NORMAL OR GAUSSIAN DISTRIBUTION (THE BELL CURVE) IS OF GREAT VALUE TO STATISTICIANS since it occurs widely throughout nature. But it is not the only distribution that does so.

In 1812, French aristocrat–cum–imperial scientist, Pierre-Simon Laplace set out the central limit theory in his *Théorie Analytique des Probabilités*. The theorem lies at the heart of statistical distributions. Events that have no duration in the normal sense—like accidents or lightning-strikes—are not contained within the normal distribution but the Poisson distribution, named for another Frenchman 25 years later. The distribution that describes events where only two outcomes are possible, like tossing a coin, is the binomial. Nevertheless, in a sense, both these distributions—and many others—are related to the Gaussian one. If one carries out several experiments or attempts to measure anything at all, the answers will not all be the same. If enough attempts are made, and the answers compared, then they will be found to be distributed in a normal distribution.

Pierre-Simon Laplace was a scientist extraordinaire, but after his death his brain was found to be smaller than average.

52 Fourier Analysis

WAVE FORMS OCCUR IN ALL TYPES OF NATURAL PHENOMENA FROM ACOUSTICS TO QUANTUM MECHANICS. The waves are usually highly complex and Fourier analysis makes it possible to describe them mathematically.

In 1822, the French mathematician Jean Baptiste Joseph Fourier, while seeking ways to describe the flow of heat mathematically, made a remarkable discovery. He believed he could prove that any wave whatsoever can be broken down into a series of sine waves, and hence that the wave can be represented mathematically by a series of simple terms. However, Fourier's approach is not quite as all-conquering as he believed. A disadvantage of Fourier analysis is that the actual calculation of the coefficients can be extremely time-consuming, even by powerful computers, but the development of the Fast Fourier Transform in the 1970s greatly accelerated such calculations, making real-time Fourier analysis a practical proposition.

Fourier analysis is an extremely powerful tool, and is key to modern signal processing, the design of musical instruments, quantum theory, spectroscopy, and much more.

53 The Mechanical Computer

A COMPUTER IN THE MODERN SENSE OF THE WORD IS A LOT MORE THAN JUST A CLEVER CALCULATOR. It is a tool that can be tasked with any job so long as it is given the correct instructions, or program. In the early 1800s, the technology still had a long way to go but it was the demand for complex math that drove it forward.

The first programmable device was not a math machine at all, but a loom invented in the 1800s by Frenchman Joseph Jacquard for weaving cloth with fine patterns. Early mechanical looms could weave faster than people, but were incapable of remembering the patterns. Jacquard's innovation was to encode the pattern as perforations on a punched card that could be read by the machine. Punched cards were still being used as a means of programming computers in the 1950s.

Turn of the wheel

While Jacquard's loom was not a computer, it did provide inspiration for English mathematician Charles Babbage, considered by many to be the father of computing, certainly the hardware. In 1822, Babbage built a prototype calculating machine which he used for testing the mechanism of a larger machine named the Difference Engine. This machine used a system of gears to calculate tables of mathematical data, but proved too pricey for Babbage to build. In the 1840s, Babbage also designed the even more complex Analytical Engine, which is considered to be the first true computer since it had a memory and was programmable.

In the 1870s, British scientist Lord Kelvin built an analog computer for calculating the tides. Instead of either being on or off like the switches of a digital computer—any state in between is impossible—the corresponding states of the cog mechanism of this device rose and fell continuously, like the tide. When turned by hand, the movements of the the dozen of so wheels traced the position of the tide onto a rotating paper drum. Kelvin's machines were so accurate the they were still in use in the 1970s.

A full size Difference Engine was built in 1991 by the Science Museum in London using Babbage's original plans.

ADA LOVELACE

Lady Ada Lovelace was the daughter of the English poet Lord Byron. In the 1840s she worked with Charles Babbage as he designed the Analytical Engine to create a punched card program for calculating Bernoulli numbers on the machine. Although that device was never built, Ada Lovelace is regarded as the first computer programmer.

54 Bessel Function

AMONG ITS MANY ROLES, MATHEMATICS IS VITAL TO THE ANALYSIS AND MODELING OF PROBLEMS IN PHYSICS, such as electricity and gravity. Pierre-Simon Laplace provided a useful tool for this in the 18th century, which was further refined by others, Friedrich Bessel chief among them.

Friedrich Bessel is also remembered for using observations of stellar parallax to calculate the distance to a star.

Much of physics deals with fields—fields of magnetism, electricity, gravitation, or fluid motion. In physics a field is a function of positions in space. In the absence of sources of the field, like point charges or masses, fields generally obey Laplace's equation, $\Delta f = 0$, where Δf is a combination of second derivatives of f. This is one of the most powerful and fundamental of the formulae used to study fields.

This is used to characterize fields in such wide-ranging applications as the way heat flows through an object, the movement of electrical current, the diffusion of gas molecules, or the motion of objects in a gravitational field. To apply the formula, Laplace's equation must be solved for each situation of interest. The solutions of Laplace's equation are called harmonic or potential functions. These were first defined by Daniel Bernoulli and further elaborated by Bessel. Bessel functions are most useful because they can be used in problems with spherical or cylindrical symmetry. They describe, for example, the vibrations of drum heads, and numerous fields of value in electromagnetism, acoustics, and hydrodynamics.

55 Group Theory

MATHEMATICS OFTEN USES TERMS DIFFERENTLY TO ORDINARY LANGUAGE, AND *GROUP* IS A GOOD EXAMPLE. In math, a group is not just a collection of things, but also a statement specifying how the group's members can be combined to make more members.

Arthur Cayley also developed group diagrams such as above, which use geometric diagrams to show the connections between a group and the set that generates it.

An example of a math group is the integers under addition, because it is always true that an integer (whole number) + another integer equals a third integer. However the integers under division is not a group, because it is not true that an integer divided by another integer always equals a third ($1 \div 2 = 0.5$ a non-integer).

Some groups, like that of integers under addition, are infinite, but others are finite—such as the numbers –1, 0, 1 under multiplication. In order to list all the members of a finite group, a Cayley table can be drawn, like this (left).

x	-1	0	1
-1	1	0	-1
0	0	0	0
1	-1	0	1

For complicated groups, Cayley tables (named for 19th-century Briton Arthur Cayley), provide a quick way of spotting patterns and properties.

Groups revealed

The rudiments of group theory were developed in the 1820s, with perhaps the first significant application being the proof by Évariste Galois (who was still at school) that some types of polynomial equation are insoluble.

Group theory is used in the analysis of symmetry. An equilateral triangle, for example, can be rotated by 120 degrees clockwise, or it can be reflected in a vertical line drawn through its center, and look the same despite these transformations. There are several more such transformations, including combinations of rotations and reflections that do alter the shape. A Cayley table can show the two subgroups of transformations (symmetrical and unsymmetrical) at a glance, but its true power is that it reveals how groups transcend simple descriptions. For example, the pattern seen in the symmetry of a triangle can occur in other areas of math that have nothing to do with triangles.

While some groups can be broken down into simpler ones, there are others which cannot be simplified any further. These are called simple groups (although the use of *simple* is perhaps another one of the anomalies in mathematical language), and their key role in symmetry has applications in quantum theory and cosmology.

The rapid development of group theory in the 1850s marked the start of a profound shift in the nature of math. Before then, equations were viewed, broadly speaking, as a shorthand for a whole series of actual calculations, with endless possible numbers in them being replaced by letters or other symbols (called variables). But with the rise of group theory, the focus shifted, making the equations and other mathematical structures things in themselves, worthy of study and development in their own right.

A manuscript from mathematician Évariste Galois shows his work on group theory written when the Frenchman was just 18. Galois' astounding career came to an end when he was killed in a duel with an artillery officer at the age of 20.

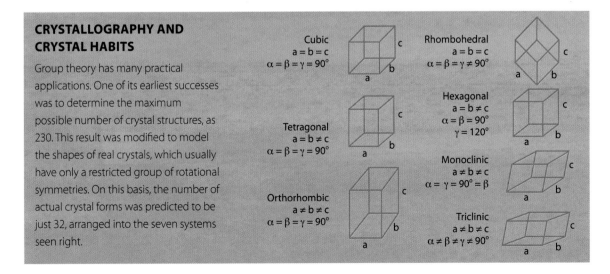

CRYSTALLOGRAPHY AND CRYSTAL HABITS

Group theory has many practical applications. One of its earliest successes was to determine the maximum possible number of crystal structures, as 230. This result was modified to model the shapes of real crystals, which usually have only a restricted group of rotational symmetries. On this basis, the number of actual crystal forms was predicted to be just 32, arranged into the seven systems seen right.

Cubic
$a = b = c$
$\alpha = \beta = \gamma = 90°$

Tetragonal
$a = b \neq c$
$\alpha = \beta = \gamma = 90°$

Orthorhombic
$a \neq b \neq c$
$\alpha = \beta = \gamma = 90°$

Rhombohedral
$a = b = c$
$\alpha = \beta = \gamma \neq 90°$

Hexagonal
$a = b \neq c$
$\alpha = \beta = 90°$
$\gamma = 120°$

Monoclinic
$a \neq b \neq c$
$\alpha = \gamma = 90° = \beta$

Triclinic
$a \neq b \neq c$
$\alpha \neq \beta \neq \gamma \neq 90°$

56 Non-Euclidean Geometry

IT IS OFTEN REMARKED THAT EUCLID'S *ELEMENTS* HAS BEEN REPRINTED ALMOST AS MANY TIMES AS THE BIBLE, AND FOR CENTURIES THIS ANCIENT GREEK WORK WAS REGARDED AS ALMOST SACRED. However, mathematicians are never complacent, and by the 19th century the nagging doubts over the inconsistencies in Euclid's assumptions led to strange new geometries, where straight lines were also curved.

Euclid's geometry is based on points, lines, and planes. (It is frequently referred to today as planar geometry.) A point has no dimensions, just a location, while a line passing through the point has just one dimension, length, but no width. The length is measured between one point and another elsewhere on the line. Euclid establishes the properties of points and lines in five postulates. The first four are uncontroversial and accepted as universal truths: I. A straight line segment connects two points. II. Any straight line can be extended indefinitely. III. A circle has a constant line segment as a radius with one end point as the center. IV. All right angles are congruent—they coincide when placed one atop the other.

Problem postulate

The fifth and final postulate is where the trouble starts: V. If two lines are drawn so they cross a third and the sum of their inner angles is less than that of two right angles, then those two lines must inevitably intersect each other if extended far enough. The implication of this is that if the angles add up to two right angles (180°) then the

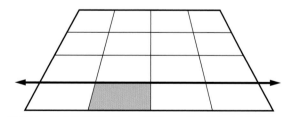

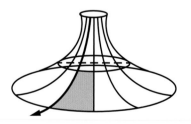

A simple comparison of the three geometries shows how a rectangle would differ when drawn on a plane, a pseudosphere (purely convex surface), and a sphere (a wholly concave one).

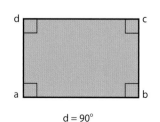

d = 90°

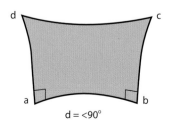

d = <90°

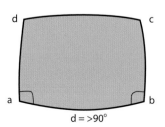

d = >90°

MANHATTAN GEOMETRY

The unbending grid of streets in a city such as Manhattan gives a certain simplicity to navigating you way around. However, the layout of parallel and perpendicular lines also creates its own geometry. What is the distance from A to B. Well, if you were a flying crow or a giant Euclid equipped with a ruler and the Pythagoras theorem, the distance measured diagonally (ignoring the road layout) is the square root of 20 (4.47 units). However, down on the street, you have to travel 6 units—whichever route you take. Using *Manhattan distances* to construct Euclidean forms give surprising results. A circle is a shape where every point is the same distance (the radius) from the center, but when the radius is measured in Manhattan distance the results are quite a surprise: the circle becomes a square!

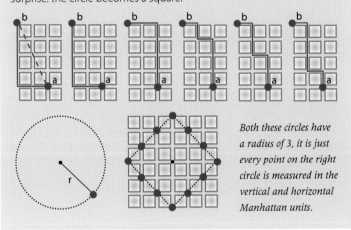

Both these circles have a radius of 3, it is just every point on the right circle is measured in the vertical and horizontal Manhattan units.

two lines will continue to infinity without intersecting—in other words they are parallel.

For centuries, the fifth postulate left a bad taste in people's mouths. Its wordiness and reliance on unstated assumptions meant it was more of a theorem than a postulate. It also appeared to be a generalized rephrasing of other sections of *Elements*, such as postulate 29—the angles of a triangle add up to 180° — and even the Pythagoras theorem itself.

Curved and straight

Generations of mathematicians wasted their careers searching for a proof of the fifth postulate, including the Hungarian Farkas Bolyai. When his son János was also lured into the problem's orbit, Bolyai despaired. He needn't have: János grew up to be a virtuoso violinist, speaker of nine languages (including Tibetan), and he was the best dancer (and swordsman) in his cavalry squadron. However, ill health drove him from miliary life and back into the arms of math and the problem of parallels.

In 1833, János Bolyai published his findings in an appendix to his father's book, *An Attempt to Introduce Studious Youth to the Elements of Pure Mathematics.* (Perhaps thankfully, the younger Bolyai's work is now just published separately with the simple title *The Appendix.*) Within Bolyai makes a breakthrough—later it became clear a similar one had been arrived at by Russian Nikolai Lobachevski a few years before: the fifth postulate was independent of the other four, and that meant it was possible to change it and create a whole new form of geometry.

Bolyai and Lobachevski suggested a geometry where the sum of the angles of a triangle is less than 180°. As in Euclidean geometry, straight lines are the shortest distance between two points, but instead of a flat plane, the lines run across a concave, or hyperbolic surface. In this hyperbolic geometry, not only are the straight lines curved, but it is possible for several parallel lines to pass through the same point.

Bending the other way

In the 1850s Bernhard Riemann began to explore a geometry based on convex, elliptical surfaces (like a sphere), where the angles of a triangle add up to more than 180°. Straight lines are curved here also, but unlike the other two geometries, they are not infinite in length but instead curve around in circles. In addition elliptical geometry does not include the concept of a parallel line.

57 The Average Person

IF YOU SAMPLE A GROUP OF PEOPLE, A FEW ARE SHORT AND A FEW ARE TALL, but most will be roughly the same height. If you plotted this variation on a graph you would get a bell-shaped curve called the normal distribution.

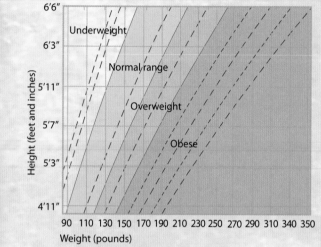

Carl Gauss had formalized the normal distribution in the early 1800s to represent characteristics that vary continuously within a group. In 1835 Adolphe Quetelet combined statistics and probability to reveal *l'homme moyen*, (the average man). His concept is applied today to plan for public health problems, and Quetelet's original study revealed human behavior in action: he took the heights of 100,000 army conscripts and compared the actual data with the expected spread. He found that the numbers of people just below and above the height limits for military service were greater than expected. Quetelet ruled out measurement errors: the conscripts were lying to avoid the draft.

The body mass index used to classify people according to an ideal weight to height ratio was originally known as the Quetelet Index.

58 The Poisson Distribution

EVENTS HAPPEN ALL AROUND US—CAR ALARMS GO OFF, PHONES RING, WEEDS GROW UP ON A LAWN—but which patterns are truly random, and which point to an underlying cause? A formula devised by the French mathematician Siméon-Denis Poisson in 1837 is a vital method of telling these possibilities apart.

The Poisson distribution deals with patterns of data occurring just by chance. Take for example the cars going along a road during the course of an hour. Knowing the hourly average does not tell us that in each minute of the hour exactly the same number of cars will go past. (In fact, if that did happen, we would suspect something funny was going on.) More likely no cars go past in one minute while two might go past the next—all by chance. Poisson's formula enables the expected number of minutes where 0, 1, 2, or more cars pass to be calculated from the overall average. Then, if the real

Siméon-Denis Poisson's many achievements earned him a place in the top 72 French scientists, whose names were inscribed on the Eiffel Tower in 1889.

data fit the predicted pattern, it's probably just chance operating; if not then there is another causal factor controlling the frequency of the cars.

Poisson's work was driven by an interest in the choices made by trial juries. It was the Polish mathematician Ladislaus Bortkiewicz who drew the world's attention to the Poisson distribution in his 1898 book *The Law of Small Numbers*.

Yes and no answers

The Poisson distribution is applicable to all walks of life: science, medicine, economics, and industry. For example, if factory machinery occasionally breaks at random, the Poisson distribution can predict the chance that several machines fail at once, allowing managers to plan accordingly.

In nature, radioactive decay is an example of a random process fitting a Poisson distribution. Comparing real data with what pure chance would predict can reveal or rule out hidden causes. For example, could cases of leukemia clustered in one location exist through chance alone or be the result of a public health hazard? Statistical analysis based on the Poisson distribution is essential for answering such questions.

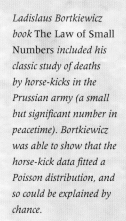

Ladislaus Bortkiewicz book The Law of Small Numbers *included his classic study of deaths by horse-kicks in the Prussian army (a small but significant number in peacetime). Bortkiewicz was able to show that the horse-kick data fitted a Poisson distribution, and so could be explained by chance.*

59 Quaternions

NUMBERS ARE NOT QUITE WHAT THEY SEEM: integers (math speak for whole numbers) are a subset of real numbers, which are a subset of complex numbers. But it does not end there. Complex numbers are a subset of *quaternions*.

Just as real numbers can be thought of as points on a number line and complex numbers as points on a plane, so quaternions can be thought of in terms of a three-dimensional mathematical space. Invented by Sir William Rowan Hamilton in 1843, they were at one time widely used in mechanics and electromagnetism but fell into disuse as simpler approaches were developed. However, they have become important once more since they are a powerful way of describing rotations in space. They are now an essential tool in computer graphics, signal processing, molecular modeling, and the mathematics of space flight.

Hamilton's greatest challenge in developing the concept was how to perform mathematical operations on quaternions—in particular, it was not clear how they could be divided by one another. When the solution struck Hamilton at last, lest he forget it, he carved it on a bridge that he and his wife were walking past in the north of Dublin.

A plaque on Dublin's Brougham Bridge commemorates William Hamilton's inspired stroll in 1843.

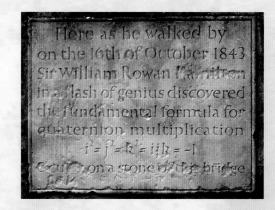

60 Transcendental Numbers

IN 1844, FRENCHMAN JOSEPH LIOUVILLE SHOWED WHAT HAD LONG BEEN SUSPECTED—the decimal representations of some numbers were not only infinitely long but also progressed in an unpredictable, patternless way.

A rational number is a number that can be expressed as a fraction, say p/q, where p and q are both integers. (Integers are the natural numbers, 1,2,3,4,5 etc, zero and the negative natural numbers.) Any number that cannot be expressed as a fraction is an irrational number. Perhaps the most familiar of the irrational numbers is π. Another is the square root of two—although Pythagoras believed that all numbers had perfect values and refused to accept the existence of irrational numbers. Transcendental numbers are a particular class of irrational numbers. A transcendental number whether real or complex cannot be expressed using algebra—in the jargon, it is not a root of a polynomial with rational coefficients.

$$0 < \left| x - \frac{p}{q} \right| < \frac{1}{q^n}$$

Liouville numbers (x), which all conform to the above expression (where p and q are positive integers), are all transcendental and provided the first evidence of such numbers.

Searching for transcendence

Proving that a number is transcendental is a very difficult thing to do. Joseph Liouville tried and failed to prove that e is a transcendental number, but did succeed in constructing an infinite class of transcendental numbers using continued fractions, and in 1851 he produced an example of a transcendental number now called Liouville's constant. This is an infinite string of 0s with a 1 positioned at every value of the exponential factorial (written $n!$) The factorial of 3 (3!), for example, is 1 x 2 x 3 = 6, so $n!$ is the factorial of every number from 1 upward. That makes Liouville's constant: 0.110001000000000000000001. In 1873 e was shown to be transcendental, while π was proven to be so in 1882. In actual fact, most numbers are transcendental—the ones with definable patterns are in the minority!

ZENO'S PARADOX

The idea that numbers did not necessarily have an exact and finite value was suggested in the 5th century BC by Greek philosopher Zeno. Zeno expressed his ideas as a number of paradoxical thought experiments based on the infinite divisibility of space. One called the dichotomy paradox says that for an object to travel a given distance it must first travel half that distance, and before that a quarter, an eighth, and so on to an infinity of divisions, with the result that the journey is never completed. Zeno returned to this concept with *Achilles and the Tortoise*, the story of a race where the athletic warrior never covers the distance covered by the reptile but just gets infinitely closer.

It was not until the development of calculus by Newton and Leibniz that a solution was found. Then it was demonstrated that an infinite geometric series can converge, so that the infinite number of "half-steps" traveled is balanced by the increasingly short amount of time needed to cross the decreasing distances.

61 Finding Neptune

THE TRUE POWER OF MATH IS DEMONSTRATED WHEN PURELY THEORETICAL CALCULATIONS RESULT IN A DISCOVERY IN THE REAL WORLD. In 1846 an illustrious French mathematician did this in style by predicting the location of a hitherto undiscovered planet.

In the 1840s the most distant planet known was Uranus, discovered 60 years before. Many decades of detailed observations of this object, only just visible with the naked eye, had revealed that its vast orbit (the planet takes 84 years to circle the Sun) did not quite conform to the path ascribed to it by Newton's laws of gravitation. The suspicion was that another, as yet undiscovered, world was pulling on Uranus, skewing its orbit.

Urbain Le Verrier (right) explains his discovery of Neptune to Louis Philippe I, the last king of France.

A race began to find it and the task fell to mathematicians. Urbain Le Verrier, stationed at the Paris observatory, solved the problem a few days before John Adams, his English rival. He used the perturbations observed in Uranus's orbit to pinpoint the next planet, and passed his findings to Johann Galle in Berlin. The German astronomer set his telescope on the planet within hours of receiving Le Verrier's letter. It looked blue, like the ocean, so the eighth planet was named Neptune, after the god of the sea.

Mistaken identity

Several years later Le Verrier suggested that the solar system contained a ninth planet, this time next to the Sun. Le Verrier's argument was that a small planet he named Vulcan lay between the Sun and Mercury, perturbing the latter planet's orbit. The Frenchman predicted that Vulcan orbited the Sun in just 19 days. For the next 50 years, scientists searched for Vulcan—all in vain, despite several mistaken sightings. In 1916, Albert Einstein used his theory of relativity to explain away Mercury's orbital anomalies. Vulcan did not exist.

62 Fechner-Weber Law

"IN ORDER THAT THE INTENSITY OF A SENSATION MAY INCREASE IN ARITHMETICAL PROGRESSION, THE STIMULUS MUST INCREASE IN GEOMETRICAL PROGRESSION." So states the Fechner-Weber Law which uses math to describe sensory perceptions and physical stimuli for hearing and other senses.

The human senses can react to an incredible range of energies. Our ears, for example, can detect sounds so quiet that the eardrum moves less than the width of a single atom, but can also hear sounds which are 10 trillion times more powerful (the limit for comfortable hearing). Similarly, the faintest star we can see is about 10 trillion times less powerful than the Sun.

When we listen to the cacophony of a busy street with traffic going by, aircraft roaring overhead, we have little difficulty in picking out quieter sounds, like conversation or the tinkle of a dropped coin. Similarly, in a piece of music, the quieter phases may well be less than one per cent as powerful as the loudest, yet, the quieter sounds are as distinguishable from each other as the louder ones.

German Gustav Fechner was a pioneer of experimental psychology. He discovered an optical illusion that carries his name, in which colors are seen on black and white pattterns. He presented his adjunct to Weber's work on sound perception in the book Elements of Psychophysics.

Enter the mathematicians

Even in the 19th century cities were far from quiet.

In terms of math, the senses respond not to the absolute increase in a stimulus but to its fractional increase from its previous level. It was Ernst Heinrich Weber who,

in 1846, discovered that the change in a person's perception of weight was proportional to the logarithm of any increase—so small increases were effectively almost imperceptible. In the case of sound, it might be that as a stimulus becomes ten times stronger, the perceived increase only doubles. In 1860, Gustav Fechner elaborated on Weber's discovery, adding his name to the concept.

One way to experience the Weber-Fechner law is to close one of two open windows; if our sense of hearing was linear with respect to intensity that would reduce the noise from outside by half. Although the sound power in the room has indeed halved, the difference in the volume we hear is barely noticeable.

63 Boolean Algebra

IN TRADITIONAL ALGEBRA, THE VARIABLES REPRESENT NUMBERS. A MACHINE USING AN ALGORITHM to solve such equations is merely a calculator. But in the 1840s, English mathematician George Boole realized that such variables could be something more.

George Boole's work culminated in the book *An Investigation of the Laws of Thought*, published in 1854, in which he proposed algebra with just two values: 1 for *true* or 0 for *false*. Instead of addition, division, and the other operations of traditional algebra, *Boolean* operations were AND, OR, and NOT, also known as conjunction, disjunction, and complement. Conjunction used the symbol ^ and worked like multiplication—any 0 in the operation resulted in a 0 (false) answer. Disjunction (v) was similar to addition but 1 v 1 is defined as 1. Finally complement (¬) is an exchange of values, 0 for 1 and vice versa. These basic operations can be expressed in a number of ways, including grids of possible outcomes, called truth tables, and these simple Venn diagrams, which show how they relate to sets of *x* and *y* (varying groups of 1s and 0s). Boole went on to derive other operations from composites of the three basic ones.

The purpose of Boole's algebra was to break down reason into its basic logical relationships and represent them with simple symbols.

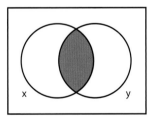

conjunction

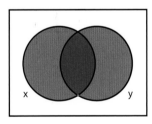

disjunction

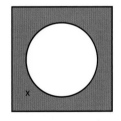

complement

In computer language

Boolean logic is nowadays associated with computer programming, especially when converted into algorithms. Although Boole's work did influence the development of mechanical computing machines of the time, his algebra was well ahead of its time, with its greatest application in computer logic taking more than a century to arrive. In the 1930s, American mathematician and electric engineer Claude Shannon was developing switching circuits and began to use Boolean equations to control when some or all of the circuit was switched on or off. These were the first logic gates, part of the foundation of digital computing. (Remember, all *digital* really means is "with numbers" and the numbers in question are the digits 1 and 0 of Boolean variables.) Basically logic gates represent the action of switches within a computer's circuits, first thermionic diodes and now millions of transistors on a single microchip. In theory a logic gate can use anything as an input, from a billiard ball to spinning photons.

64 Maxwell-Boltzmann

THE MAXWELL-BOLTZMANN DISTRIBUTION IS THE WORK OF TWO 19TH-CENTURY PHYSICISTS, JAMES CLERK MAXWELL AND LUDWIG BOLTZMANN, who worked independently to create the area of physics known as statistical mechanics, the first application of statistics to the physical sciences.

In the mid-19th century, thermodynamics (the physics of heat) was already well established, but the idea that all matter was made up of atoms was still considered suspect by some scientists. Unhappily for Boltzmann, who worked throughout his life in statistical mechanics—which assumes that matter is made up of particles—the ongoing debate led to disputes, and may ultimately have contributed to his suicide in 1906. A few years later, the reality of atoms was not only accepted, but scientists were beginning to probe their inner structure.

The idea that heat is the movement of the particles that make up matter had been proposed over a century before Maxwell and Boltzmann, by Swiss mathematician Daniel Bernoulli. According to the kinetic theory of gases, the constant motion of gas particles underlies the macroscopic qualities of gases, such as heat and pressure, that thermodynamics deals with.

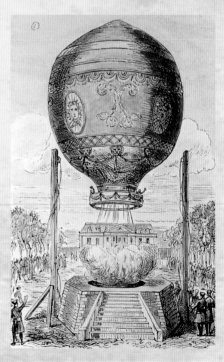

Montgolfier's pioneering hot-air balloon flight of 1783 was only fully explained when the Maxwell-Boltzmann distribution described how heating the air resulted in the gas within reducing in density as its volume went up, due to an increase in the velocity of its particles.

As well as contributing to the mathematical tools that describe the motion of gas molecules, James Clerk Maxwell was also instrumental in figuring out the math of electromagnetism.

Fluid situation

In 1859 Maxwell proposed a law to describe the velocities of particles in a gas. Since there are too many particles to be described individually, a statistical law was needed. Some scientists thought that the collisions between particles would make all the velocities equal out, but Maxwell argued that there would be a range of velocities. The following year, Maxwell's theory correctly predicted that the viscosity of a gas does not depend on its pressure. Maxwell's work inspired Boltzmann, then a student in Vienna, and in 1871 he published a more general version of Maxwell's law in terms of the distribution of energy, rather than velocity. This has become known as the Maxwell-Boltzmann distribution.

65 Defining Irrationals

IRRATIONAL NUMBERS WERE FAMOUSLY DISCOVERED BY THE PYTHAGOREANS IN ANCIENT GREECE—dealing a blow to their belief that numbers and the ratios between them were the origin of all things. These inconvenient magnitudes became confined within geometry and exiled from arithmetic —a situation that was only remedied in the 19th century.

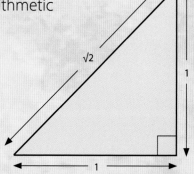

As the Pythagoreans had discovered, some magnitudes are "incommensurable" in relation to one another. In other words, the numbers cannot be expressed as a ratio of one number to another. This meant that, prior to decimal notation, formalized in the 16th century, there was no way to express these quantities numerically. Even with decimal notation they cannot be expressed precisely because the decimal digits never end.

Irrational numbers include the length of the diagonal of a square in relation to its sides (i.e. the square root of two; above) and the length of the side of one cube to the side of another cube that is double the volume of the first—the cube root of two.

Making the cut

In the 4th century BC Eudoxus dealt with this problem with a sophisticated definition of which magnitudes are comparable and which are not. Some historians see this work, included in Euclid's *Elements*, as equivalent to the current best definitions arrived at in the 1870s by German mathematician Richard Dedekind.

A definition of irrationals was needed to give calculus a full basis in arithmetic rather than a partial geometric one, as it had been left by its 17th-century inventors. Dedekind complained, "The statement is so frequently made that the differential calculus deals with continuous magnitude; and yet an explanation of this continuity is nowhere given." Continuity cannot be provided by rational numbers alone. On a number line, every rational number is represented by a point, although it is not true that every point corresponds to a rational number—some are irrational. Points representing only rational numbers would have gaps.

The intervals between the frets on a guitar, like other musical scales, are positioned according to approximations of an irrational number, the twelfth root of 2.

Dedekind's definition included all *real* numbers—both rational and irrational. He introduced what is now known as a *Dedekind cut*, which divides the number line into two parts: any real number can be identified with a cut that separates the numbers into sets of those greater than and those less than that number. For example, the square root of two is the cut that separates numbers with squares that are less than two from those with squares more than two.

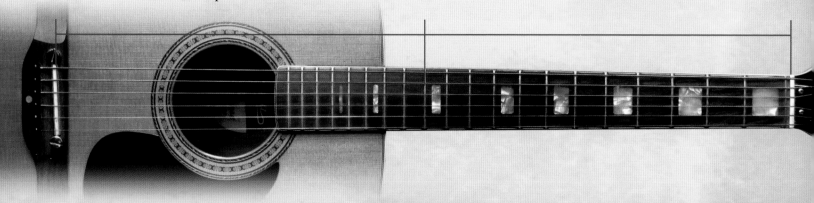

66 Infinity

To the ancients, infinity was a taboo subject, best left in the lap of the gods and also by the best definitions of the day indescribable, which played havoc with any math-based rationality. It is hard to imagine their responses to Georg Cantor's 1874 revelations on the subject—there was more than one type of infinity, and some were bigger than others.

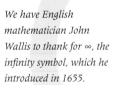

It is often thought that infinity is a very large number, but that is not really the best way to think about it. No matter how big a number is, it cannot approach infinity. Take the googol for example, the nonsense name given to the number 10^{100}—a ten followed by 100 zeros. That is a pretty large value—a billion has a mere 9 zeros, a trillion 12, etc. The number was named by nine-year-old Milton Kasner, the nephew of Edward Kasner who turned to the child for inspiration in 1920. It is of course the inspiration behind the company name Google. The internet search giant is headquartered at the Googleplex in San Jose. However, a googolplex is a different number. This time 10^{googol}, of 10 with a googol of zeros. Carl Sagan, the great science communicator, said that a piece of paper large enough to write out a googolplex would not fit in the known Universe, "and yet it is precisely as far from infinity as the number one."

CARDINAL NUMBERS

Cantor needed to specify the difference between the various types of infinity he had discovered. He accomplished this through the concept of cardinality. Cardinality classifies sets of numbers solely by their extent, so the sets (1,2,3,4,5,6,7), and (red, orange, yellow, green, indigo, violet) both have cardinality 7, as can be confirmed by the fact that the members of each can be mapped one-to-one onto the other.

The one-to-one test of cardinality can be applied to infinite sets too; the smallest type of infinity is one that is countable, like the natural numbers: 0, 1, 2, 3. Although one could never come to an end of counting them, it is at least obvious how one would start the job. Such infinities are labeled with the cardinal number $\aleph_0$ (aleph-zero). There is a whole range of cardinals for different infinities.

Everything and nothing

The concept of infinity is filled with paradoxes, and remained very much in the realm of philosophy as people struggled to imagine things that were infinitely large—and infinitely small. For example, John Wallis, the inventor of the infinity sign ∞ believed that negative numbers were not less than zero (he felt that was impossible) but just infinitesimally small.

The great German math professor David Hilbert would explain infinity to his students by asking them to imagine they are the receptionist at a hotel with an infinite number of rooms. There comes a day when Hilbert's Hotel is fully booked, but the staff never turn anyone away. When one more guest arrives, the receptionist calls his infinity of guests, asking them all to move to the room with the number one

We have English mathematician John Wallis to thank for ∞, the infinity symbol, which he introduced in 1655.

This 18th century text by Pierre Raymond de Montmort is concerned with infinite series, sequences of numbers derived by a particular operation but with no defined end point.

more than their current one, and the new guest now takes room number 1. However, just as he is settling in, an infinity of tourists arrive all at once. The receptionist is unperturbed, asking his current guests to move to the room with a number double that of their current one—and in so doing, an infinity of rooms becomes available.

Cantor's infinities

Hilbert's hotel has often been a prelude to a lecture on the work of fellow German Georg Cantor, who in the 1870s put infinity on a firm mathematical footing. Indeed he found that there was not just one infinity but—as might be expected—an infinity of them.

The first infinity is the one we all understand, that of the natural numbers, or counting number 1,2,3... etc. Just like the numbers on the door of Hilbert's Hotel, you can keep on adding one forever. Cantor described this infinite set of numbers as *countable*, meaning it is possible to count them (if you had an infinite amount of time.) The natural numbers are all rational—meaning they can be expressed as a fraction. However, most fractions are not natural numbers. Nevertheless, although natural numbers are a subset of rational numbers, both sets are infinite and countable. In the terms Cantor used, they have the same *cardinality*.

Getting real

Cantor would have been familiar with Richard Dedekind's work, a decade earlier, on irrational numbers. Therefore he knew that the infinity of rational numbers was just a subset of real numbers, the rest of the real numbers being the irrational numbers that could not be expressed as fractions, just infinitely long, patternless strings of decimals. Unlike the rational infinity, the full set of real numbers could not be counted—even if you wanted to and had the time—and so Cantor described this type of infinity as being *uncountable*, and had a cardinality greater than the countable infinities.

But the story does not stop there. Cantor identified subsets within real numbers. Some irrational numbers (and all rational ones) are what Euclid described as being constructible—they could be arrived at using a geometric process. For example, the highly irrational $\sqrt{2}$ is constructible by drawing the diagonal of a square. Cantor showed that the constructible numbers were a countable infinity. All constructible numbers are algebraic—they can be expressed with algebra, and the infinity of algebraic numbers is also a countable infinity. However, the non-algebraic irrationals, or transcendental numbers are uncountable in their infiniteness. We will stop now, this could go on forever.

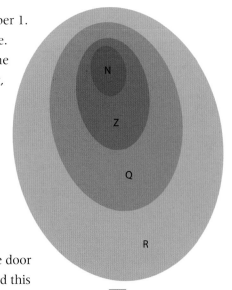

■ Countable infinities
■ Uncountable infinities

N is the infinite set of natural numbers; Z is the set of whole numbers (including negative ones); Q includes all the rational numbers; R is the set of real numbers.

Cantor's pairing proof shows that the infinite set of fractions has the same cardinality as the infinite set of natural numbers.

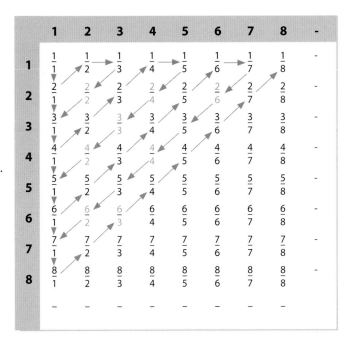

	1	2	3	4	5	6	7	8	-
1	$\frac{1}{1}$	$\frac{1}{2}$	$\frac{1}{3}$	$\frac{1}{4}$	$\frac{1}{5}$	$\frac{1}{6}$	$\frac{1}{7}$	$\frac{1}{8}$	-
2	$\frac{2}{1}$	$\frac{2}{2}$	$\frac{2}{3}$	$\frac{2}{4}$	$\frac{2}{5}$	$\frac{2}{6}$	$\frac{2}{7}$	$\frac{2}{8}$	-
3	$\frac{3}{1}$	$\frac{3}{2}$	$\frac{3}{3}$	$\frac{3}{4}$	$\frac{3}{5}$	$\frac{3}{6}$	$\frac{3}{7}$	$\frac{3}{8}$	-
4	$\frac{4}{1}$	$\frac{4}{2}$	$\frac{4}{3}$	$\frac{4}{4}$	$\frac{4}{5}$	$\frac{4}{6}$	$\frac{4}{7}$	$\frac{4}{8}$	-
5	$\frac{5}{1}$	$\frac{5}{2}$	$\frac{5}{3}$	$\frac{5}{4}$	$\frac{5}{5}$	$\frac{5}{6}$	$\frac{5}{7}$	$\frac{5}{8}$	-
6	$\frac{6}{1}$	$\frac{6}{2}$	$\frac{6}{3}$	$\frac{6}{4}$	$\frac{6}{5}$	$\frac{6}{6}$	$\frac{6}{7}$	$\frac{6}{8}$	-
7	$\frac{7}{1}$	$\frac{7}{2}$	$\frac{7}{3}$	$\frac{7}{4}$	$\frac{7}{5}$	$\frac{7}{6}$	$\frac{7}{7}$	$\frac{7}{8}$	-
8	$\frac{8}{1}$	$\frac{8}{2}$	$\frac{8}{3}$	$\frac{8}{4}$	$\frac{8}{5}$	$\frac{8}{6}$	$\frac{8}{7}$	$\frac{8}{8}$	-
	-	-	-	-	-	-	-	-	-

67 Set Theory

ONE OF THE MOST UBIQUITOUS CONCEPTS IN MODERN MATHEMATICS IS THAT OF THE SET. Anything can be classified as a member of one or more sets, and set theory is concerned with the way sets and their subsets are related, not least in the way they can or cannot be converted from one to another.

The number 4 is a member of more than one set, including the sets of integers, of square numbers, even numbers, numbers with an English name that contains the same number of letters as they represent, numbers less than 20..., the list goes on. The set of even numbers is an example of a subset—every member is also in the set of integers. The way sets intersect, forming subsets, is perhaps most familiar in the form of Venn diagrams, invented by British logician John Venn in 1880 during the early days of set theory.

    However, the power of sets is not simply that of classification, it is also a way of capturing the concept of a function. A function, roughly speaking, is a way in which a number is modified. Apply a function to a number and you get another one: square 4 and you get 16, so squaring is a function; apply that function to the set {-1, 0, 1} and we get the set {1,0,1}. The squaring function is written $f(x)=x^2$.

Georg Cantor is the founding figure of set theory, although his ground-breaking work is nevertheless now known as naive set theory because of paradoxical flaws later found in it.

Beginnings and no end

The combination of sets and functions led to far-reaching and fundamental changes in mathematics, as part of Georg Cantor's work into infinity in the 1870s. Basic to his vision was the point that some sets are finite, like "the integers from 1 to 7," and some are infinite, such as "the integers." In some cases there is a one-to-one correspondence between different sets, such as between the set of "official" (according to Newton at least) colors of the rainbow and the set of integers from 1 to 7.

    Any two different finite sets may or may not have this one-to-one correspondence (that is, the same number of members). Cantor's seemingly innocent question was, "What about infinite sets? Do they necessarily have one-to-one correspondence?" The surprising answer was no. The set of integers is infinite, and yet between any two integers there is a limitless number of real numbers (which include 0.1, 1/3, √2 and π). So there is no one-to-one correspondence between both

This Venn diagram shows six sets all intersecting on another.

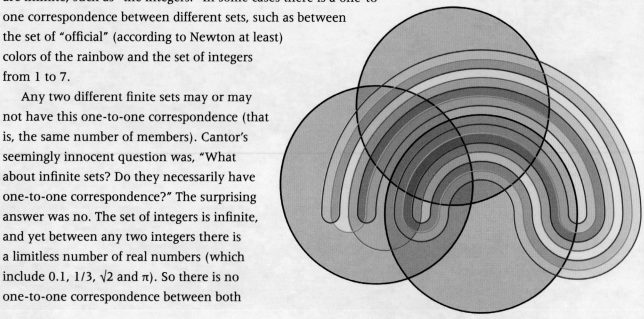

$$\text{let } R = \{x \mid x \notin x\}, \text{ then } R \in R \Leftrightarrow R \notin R$$

sets, which means that the infinite set of transcendental numbers is in some sense larger than the infinite set of integers.

Russell's Paradox, an inherent contradiction in set theory, is represented by this expression, here R is the set of all sets.

A flaw within

The power of set theory, both within and outside mathematics, is vast, so when it was challenged by Briton Bertrand Russell in 1901, a major battle was at hand. Russell's preliminary query was, just like Cantor's, deceptively simple. Before we get to it we first need to follow his reasoning:

1. Some sets are members of themselves, such as "the set of sets."
2. Most sets are not members of themselves. "The set of integers" is not an integer and is therefore not in the set.
3. Imagine listing all the sets that are not members of themselves. Let's call this big list of sets, A.

Russell's question was: "Is set A a member of itself?" Let's say yes, A is a member of A. But that can't be: by definition, A is a set of sets that are not members of themselves, and so A cannot be a member of itself. So, let's say no, A is not a member of A. But that can't be true either: "a set that is not a member of itself" is what defines membership of A, so A must be in it!

In ordinary language, Russell's paradox sounds more like a riddle. It has several forms, and one is as follows: there is a village in which all the men either shave themselves or are shaved by the village barber. In the language of sets one would say that the whole population of the village is therefore in one of two sets: "men who shave themselves" or "men who are shaved by another." Russell's question becomes, "Which set is the barber in? He can't be in the first set because he is the barber and only shaves men who do not shave themselves. So he can't shave himself. So he must be in the second set, so the barber must shave him. But that's not right either—he does not shave himself."

This paradox was a death blow to Cantor's set theory, but fortunately, a better set theory arose from its ashes. Today's set theory avoids paradoxes like that of the hirsute barber by making it a rule that sets cannot contain themselves and also denying the possibility that the collection of all things counts as a set, thus declaring the paradox is against the rules.

SIERPINSKI CARPET

Among many other applications, sets are used to define shapes and their properties. A set named after Cantor because he developed it in 1883, although he was not the first to define it, is one with some surprising properties. The members of the Cantor set are points on the segment of a line. The set can be extended to describe flat surfaces too, and the pattern (below) that results is called the Sierpinski carpet (after Pole Wacław Sierpiński). This pattern, first defined in 1916, is one of the first examples of a fractal, since it is made up of endlessly repeating self-similar patterns.

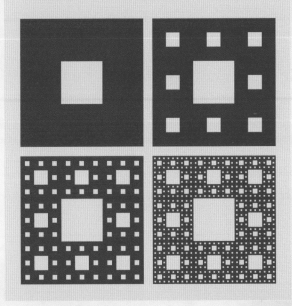

68 Peano Axioms

PUBLISHED IN 1889, A SET OF STATEMENTS KNOWN AS THE PEANO AXIOMS ARE FUNDAMENTAL TO MATHEMATICS. They set out the assumptions needed to establish the existence of the natural numbers.

It is perhaps surprising that arithmetic, simple operations such as addition and subtraction were not entirely formalized until the late 19th century. The axioms of Italian Giuseppe Peano did for arithmetic what Euclid's assumptions did for plane geometry, breaking it down into its simplest concepts. The nine axioms begin by establishing the first natural number. Initially Peano chose one for this but later changed it to zero. Several of the other axioms use the idea of *successor* numbers, showing that they apply to all natural numbers in succession. The axioms have been largely unchanged since Peano proposed them, although there has been some rewording to make the mathematical logic more robust.

Peano was also interested throughout his life in language and notation, and introduced several new math symbols. His papers were so full of symbols—and free from words—that some commentators complained that they looked like wallpaper!

1. 0 is a natural number. The next four axioms describe how numbers relate to each other.
2. For every natural number x, $x = x$.
3. For all natural numbers x and y, if $x = y$, then $y = x$.
4. For all natural numbers x, y and z, if $x = y$ and $y = z$, then $x = z$.
5. For all a and b, if a is a natural number and $a = b$, then b is also a natural number. That is, the set of natural numbers are closed under the previous axioms.

Four more axioms describe the arithmetical properties of natural numbers.

69 Simple Lie Groups

IN 1888, A PAPER WRITTEN BY WILHELM KILLING, A GERMAN MATHEMATICIAN, PROPOSED A MAJOR NEW PROJECT—the full classification of simple Lie groups.

Perhaps a stranger to understatement, quantum physicist A.J. Coleman declared that Killing's proposal was the "greatest mathematical paper of all time." An example of a Lie group (pronounced lee and named for Norwegian Sophus Lie) is the symmetry group of a circle: the set of rotations and reflections that can be applied to a circle and which leave it looking unchanged. Symmetries pertain to other shapes, including those in other dimensions, which are known as manifolds. Lie groups describe the symmetries of manifolds.

The *simple* Lie groups are those that cannot be broken down further. They are used to describe the particles that carry the forces of nature: gravitation, electromagnetism, the strong force that holds atomic nuclei together, and the weak force that is a factor in radioactivity.

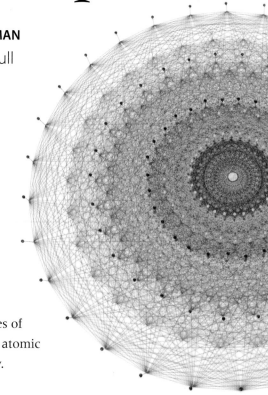

This 6,720-edge shape with 240 vertices (corners) is derived from the E_8 Lie group, which is being used to investigate string theory.

70 Statistical Techniques

THE WORD *STATISTICS* HAS TWO MEANINGS. It refers to collections of data that can be presented as tables or graphs, but it also means the science of developing mathematical tests for analyzing this data.

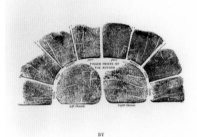

FINGER PRINTS

BY
FRANCIS GALTON, F.R.S., ETC.

Francis Galton had a colorful career including discovering that a person's fingerprints were unique and could be used to identify them.

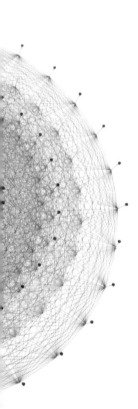

As society industrialized and populations exploded in the 19th century, the need to quantify the world also grew. Sometimes the figures "spoke for themselves," but often the important information was hidden away and had to be unearthed. Three British scientists were prominent in the development of statistical techniques. Francis Galton, a wealthy cousin of Charles Darwin, became fascinated by trying to measure how physical and mental features of humans might be inherited. Galton pioneered new methods of analyzing data, and in 1888 published his groundbreaking work on *correlation*—measuring the apparent relationship between two varying quantities. Galton loved measuring things: among his topics were whether prayer worked, and how beauty varied in different places. (The women of Aberdeen, Scotland, were the least beautiful, apparently!)

Galton also discovered the *wisdom of crowds* where a collective prediction from a large group of people produces an average that often turns out to be highly accurate. For example, Galton found that the average of 800 guesses taken during a village fair of the weight of a slaughtered cow was closer to the true value than any one of the guesses.

Later developments

Inspired by Galton, Karl Pearson put statistics on a sounder mathematical footing, introducing in 1900 his famous chi-square test to check how well actual data fitted a theoretical curve. (Pearson and Galton were also both enthusiasts for eugenics—the practise of improving the human race through selective breeding. Eugenics attracted many followers including the Nazis, but was eventually deemed immoral.)

Further advances were made by R.A. Fisher, who developed a means to analyze the variance of a sample—revealing if a sample size was large enough to be meaningful. In the days before computers, Fisher also designed his tests so they would not involve a massive amount of number crunching by researchers.

THE LADY WITH THE STATS

Florence Nightingale is remembered as a nurse—the founder of the profession no less. However, her success was due to her skills as a statistician. She collected mortality data from the hellish Selimiye military hospital on the edge of Istanbul during the Crimean War and then was able to show that her regime of hygiene prevented many of the deaths. Her presentation to the top brass back in London presented statistics in *rose diagram*, which Nightingale had invented. These very visual aids have since been renamed pie charts.

Florence Nightingale at work at Selimiye.

MODERN MATHEMATICS

71 Topology

THE MATHEMATICS BEHIND LOCATION AND DISTORTION IS KNOWN AS TOPOLOGY, ALSO KNOWN AS "BENDY GEOMETRY." The field has its origins with Leonhard Euler in the 18th century, but as early as 1676, Gottfried Leibniz had called for "a new geometry of position".

In topology, as opposed to classical geometry, the arrangement of shapes is all-important. The student passing through the doors to this mind-warping domain must forget about distance, angle, and measurement—all prerequisites for Euclidian geometry—instead the relative positions of shapes, their connectedness, and organization matter more.

Topology grew out of the graph theory that began with Euler's 1736 classic paper, the *Seven Bridges of Königsberg*. This work is inspired by a puzzle based on the layout of the Baltic port—was it possible to find a route that crossed all seven of the city's bridges just once? Many had tried, and almost everyone believed that it was impossible. Euler's interest was why. He correctly ascertained that the significant factor was the number of bridges, or connections, not the distances and orientations. He then showed that bridge problem has no solution—you had to miss out one bridge or cross another twice.

A Klein bottle, first described by Felix Klein in 1882, is a non-orientable two-dimensional surface. Like a Möbius strip it has just one face. However, unlike the strip, to form one by deforming the surface requires moving it through a fourth spatial dimension. There are of course only three of those.

Frenchman Henri Poincaré's investigations into the "connectedness" of topological spaces in the late 19th century led to a famously unfathomable conjecture that was finally solved in 2002 by the Russian mathematician Grigori Perelman.

Topological equivalence

Taking their cue from Euler's bridge problem topologists turn every shape into a network of nodes and connections—the features of a shape that remain the same no matter how the faces are distorted. Even vastly different shapes can actually be the same to a topologist. In the jargon the shapes have *topological equivalence*.

The test of how two shapes are related is whether they can be persuaded to morph into one another. It is easy to see how a football and soccer ball are related—the egg-shaped one will turn into a sphere if inflated. On a more macabre note, if you were inflated, would you turn into a ball? Luckily no, because your alimentary canal makes you a torus—a donut.

From the point of view of topology, a donut and a coffee mug are the same shape. Both have a hole going through them, and the differences in the distortion of the surfaces of either shape is incidental.

Problems, real and unreal

More formally, notions of homology and homotopy, both of which rely heavily on set theory, govern the equivalence of shapes. Homology examines the holes in shapes and is the more intuitive. Homotopy is more abstruse, dealing with information that spaces contain, and how their functions can be deformed continuously. Both retain that hallmark of the field—general results can be inferred from qualitative, rather than quantitative, investigations.

Other branches of topology investigate how geometric functions can become entangled and disentangled, looking into the geometry of knots and multi-dimensional surfaces called manifolds. Despite the impossible tricks that topologists pull—things like turning spheres inside out—topology has many real-world applications. Its ability to recognize classes of objects without needing to know precise shapes or measurements makes it a formidable system. Cooperative swarmbots—small autonomous robots that work en masse—use topological spaces to keep track of their surroundings, and the math is used to figure out where to place cell phone masts to give ideal network coverage. Geographical Information Systems (GIS) class the elements of maps as topological domains and boundaries, allowing users to work out relationships between real-world objects, while eliminating errors introduced by measurement and scaling.

MÖBIUS STRIP

The Möbius strip is a unique surface that has just one side and one edge. With a sheet of paper, scissors, and glue or sticky tape, you can enter this strange dimension of geometry. Take a strip of paper and join the ends, but before closing the loop put a single twist in it. This endless loop is known as a non-orientable surface, because it has no inside or outside. If we assume that the paper has zero thickness, the manifold has only one face. If an ant were to crawl around the loop he would cover the whole surface and end up on the opposite side to the one on which he began.

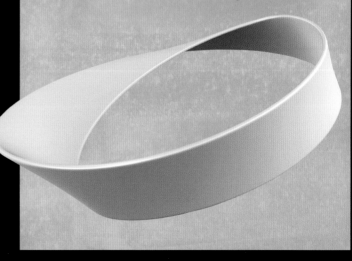

72 A New Geometry

AS THE 19TH CENTURY DREW TO A CLOSE, DAVID HILBERT PROPOSED THE UNTHINKABLE: he was going to replace Euclid's geometry, a central plank of mathematics.

For over 2,000 years, Euclid's geometry was regarded as a masterpiece. However, as time passed, more and more weaknesses were detected in his work. Many of his definitions were not really that clear and required more assumptions than those stated by Euclid. Such criticisms came to a head with David Hilbert's 1899 *Foundations of Geometry*. Hilbert identified Euclid's undoing as the fact that the ancient Greek's axioms or assumptions were drawn from the characteristics of the real world: real points, real lines, real curves, and real shapes. The errors in those assumptions were therefore hidden because the math worked superficially, although in mathematical reality the axioms were not fully defined and were mathematically incorrect. Euclid's diagrams exacerbated this problem, since they sometimes made things that were not actually known to be true appear to be true. For instance, putting a point on a straight line means that this point (2) lies between points 1 and 3 that are the end points of the line. However, the concept of "betweenness" requires a much more precise definition than that.

Hilbert's response was to take an approach opposite to Euclid's. Rather than geometry being a way of exploring the properties of shapes, points, and lines, it was to be a subject about the logical relations of symbols, irrespective of whether those symbols represented lines or anything else—or nothing at all. This new, formalist approach was applied by Hilbert and many others to more areas of mathematics and was soon to become the standard practice.

As might be expected of the man who did away with Euclid, German David Hilbert was one of the most influential mathematicians of all time. His influence is seen in topology and the philosophy of math.

73 Hilbert's 23 Problems

AT THE FIRST INTERNATIONAL CONGRESS OF MATHEMATICIANS OF THE 20TH CENTURY, DAVID HILBERT SET A TASK FOR HIS COLLEAGUES, FUTURE AND PRESENT. At the Paris conference he set out the ten thorniest unsolved problems that would occupy math for the coming century and circulated the details of 13 other areas of investigation. Hilbert chose his challenges in the aim of the furthering of mathematics. At the turn of the following century, only three remained unresolved.

23 PROBLEMS

1 Continuum problem as described by Cantor. Is there a transfinite number between that of a countable infinity set (such as the integers) and the numbers of the continuum? The continuum hypothesis as developed by Gödel and Cohen suggests that the answer depends on the particular version of set theory assumed and was related to Zermelo's axiom of choice of 1904. In 1963, Cohen showed that the axiom of choice (being able to select a single, precise value from an infinite set) was independent of the other axioms in the theory. However not everyone agrees that this problem is solved.

2 To establish the compatibility of the arithmetical axioms. Russell and Whitehead attempted this, but Gödel's incompleteness theorem shows that it can never be proved that the axioms of logic are consistent. Any system that can formulate its own consistency can prove its own consistency even if it is inconsistent. There is no consensus whether Gödel's work is a solution to this problem.

3 Can two tetrahedra (or other polyhedra) of equal volume always be decomposed (broken up and rearranged) into one another? Within weeks of Hilbert's announcement, Max Dehn showed that a regular polyhedron cannot always be decomposed into a finite number of smaller but congruent (all the same size) polyhedra.

4 Find geometries whose axioms are closest to those of Euclid's if the ordering and incidence axioms are retained. To do this the congruence axioms (that make shapes the *same* or *different*) have to be expanded and the parallel postulate omitted. Although Georg Hamel, a student of Hilbert, suggested a solution, few mathematicians class this as a valid problem, merely a vague pondering.

5 A generalization of the Cauchy functional equation—are continuous Lie groups also differential groups? This problem is only partially resolved.

6 Can physics—and in so doing all science—be broken down into groups of fundamental axioms as math has been? The answer: maybe, but not quite proven yet.

7 If a is an algebraic number other than 0 or 1 and b is irrational is a^b then transcendental? If b is also algebraic then the answer is yes. If b is non-algebraic—i.e. transcendental itself—the problem is still unresolved.

8 Prove the Riemann hypothesis: the real part of any non-trivial zero of the Riemann zeta function is half. This problem remains open and is now on the list of problems for the 21st-century's mathematicians to solve.

9 Construct a generalization of the reciprocity theorem for algebraic numbers. Partially solved.

10 Is there a universal algorithm for solving Diophantine equations? In short, no. In the 1970s Yuri Matiyasevich used the Fibonacci sequence to show that the solutions grow exponentially indicating that problem 10 is impossible.

11 Extend the results obtained for quadratic fields to arbitrary integer algebraic fields. Partially resolved.

12 Extend the Weber-Kronecker theorem from rational numbers to any algebraic fields. Ring field theory partially resolves this problem.

13 Show that it is impossible to solve a generalized seventh-degree equation with functions using two variables. As yet only partially resolved.

14 Does an algebraic group acting on a polynomial ring always generate a finite ring of invariants? Definitely not.

15 Provide a rigorous basis for Schubert's calculus. Progress has been made but still not fully resolved.

16 Study the topology of real algebraic surfaces. More an area for study than a soluble problem.

17 Find a representation of definite rational functions using sums of squares. Resolved in 1927 when Artin and Delzell found there was an upper limit. A lower limit is still being sought.

18 Build spaces with congruent polyhedra to find closest packing of spheres and three-dimensional anisohedral tiles. Resolved in 1998.

19 Are the solutions to variational problems always analytic? Resolved in 1957.

20 Solve general boundary value problems.

21 Do linear differential equations conform to a monodromy group? In some cases but in general no, according to Bolibruch in 1989.

22 Find uniformity in analytic functions. Resolved.

23 Extend the methods of calculus of variations. As yet ongoing.

D. HILBERT. — PROBLÈMES MATHÉMATIQUES. 79

comme un système de transformations

$$x'_i = f_i(x_1, \ldots, x_n; a_1, \ldots, a_r) \quad (i = 1, \ldots, n)$$

tel que deux transformations quelconques

$$x'_i = f_i(x_1, \ldots, x_n; a_1, \ldots, a_r),$$
$$x''_i = f_i(x'_1, \ldots, x'_n; b_1, \ldots, b_r)$$

du système, opérées l'une après l'autre, fournissent une transformation appartenant également au système et, par suite, représentable sous la forme

$$\ldots, c_r),$$

SUR LES

PROBLÈMES FUTURS DES MATHÉMATIQUES,

Par M. David HILBERT (Göttingen),

TRADUITE PAR M. L. LAUGEL (¹).

Qui ne soulèverait volontiers le voile qui nous cache l'avenir afin de jeter un coup d'œil sur les progrès de notre Science et les secrets de son développement ultérieur durant les siècles futurs? Dans ce champ si fécond et si vaste de la Science mathématique, quels seront les buts particuliers que tenteront d'atteindre les guides de la pensée mathématique des générations futures? Quelles seront, dans ce champ, les nouvelles vérités et les nouvelles méthodes découvertes par le siècle qui commence?

L'histoire enseigne la continuité du développement de la Science. Nous savons que chaque époque a ses problèmes que l'époque suivante résout, ou laisse de côté comme stériles, en les remplaçant par d'autres. Si nous désirons nous figurer le développement présumable de la Science mathématique dans un avenir prochain, nous devons repasser dans notre esprit les questions pendantes et porter notre attention sur les problèmes posés actuellement et dont nous attendons de l'avenir la résolution. Le moment présent, au seuil du vingtième siècle, me semble bien choisi pour passer en revue ces problèmes; en effet, les grandes divisions du

(¹) L'original de la traduction a paru en allemand dans les *Göttinger Nachrichten*, 1900. M. Hilbert a fait ici quelques modifications à l'original au § 13 et quelques additions au § 14 et au § 22.

(L. L.)

74 Mass Energy

IN THE 19TH CENTURY IT WAS THOUGHT THAT GRAVITY POWERED THE SUN, THAT THE HUGE CLOUD OF GAS AND DUST from which the star formed got hotter as gravity made it contract. But then Einstein said it was all down to $E=mc^2$, perhaps the most famous math equation of all.

The gravity theory of solar energy fell apart when it was calculated that if the Sun shrank by around 50 meters per century it would burn for just 100 million years. The trouble with this theory was that it conflicted with geological evidence that the Earth was billions of years old. The solution came with a flash of genius. In 1905 Albert Einstein produced his theory of special relativity, overturning our understanding of space, time, matter, and energy. One of his conclusions was that the speed of light is the ultimate speed bump—it is a limit that may be approached but never passed. As a consequence of this, we are led to that most famous of equations $E=mc^2$.

$$E=mc^2$$

This equation hails from Einstein's 1905 theory of special relativity. It states that energy (E) and mass (m) are proportional, with the constant relating them (c) being the square of the speed of light.

ATOMIC BOMB

Mass is converted into energy by nuclear fission, which powers the explosions of nuclear weapons. The atomic bombs dropped on Japan at the end of World War II each converted about half a gram of nuclear material into energy, creating explosions large enough to devastate entire cities.

Massive discovery

$E=mc^2$ says that energy and mass are equivalent, and a small amount of mass equals a tremendous amount of energy. The power source of the Sun is nuclear fusion: four hydrogen atoms fuse to form one helium atom. However, the mass of one helium atom is less than the sum of four hydrogens. A simple calculation tells us that every second 4.2 billion kilograms of the Sun's mass is converted into energy.

The precision of Einstein's equation was revealed in 2008. We now know that protons and neutrons in atoms are themselves formed from still smaller particles named quarks. However, the quarks accounted for only about five per cent of an atom's mass. Where was the rest? Supercomputer calculations at France's Centre for Theoretical Physics showed that the missing mass was accounted for by the energy associated with the motion and interactions of the subatomic particles— a vindication of Einstein on a subatomic scale.

75 Markov Chains

BASED ON THE WORK OF THE RUSSIAN ANDREY MARKOV IN 1907, MARKOV CHAINS ARE STATISTICAL MODELS, which have found a great number of applications in information theory.

Markov began his career as a financial manager for a Russian princess.

The term *chain* refers to a process set to run for a specific time or predetermined number of cycles. Each step of the cycle is random and "memoryless," which means that every step is dependent only on the one before it and not on events further in the past, unlike a coin or dice toss, in which every throw is independent of the previous one.

With their sensitivity and dependence on initial conditions, Markov chains generate highly random outcomes and it is impossible to predict the state at any given point during the cycle. Nevertheless, the overall statistical properties of the system are predictable, which makes them good models for real life. Markov chains are used to model a wide range of physical phenomena, such as market prices, enzyme activity, the evolution of chemical systems, and in Google's famous PageRank formula, which provides a measure of the importance of a web page somewhere on the Internet.

76 Population Genetics

IN 1908 THE NEW SCIENCE OF GENETICS WAS IN A MUDDLE. THE CONCEPT OF THE GENE WAS EMERGING, but people couldn't see how genes spread through populations. Fortunately, mathematics came to the rescue.

This graph of the Hardy-Weinberg Principle shows how the frequency of genotypes (aa; Aa; and AA) relate to the frequency of alleles (p=A; q=a) in a population.

Every gene has several versions, or *alleles*. Our cells have two alleles for every gene, one from each parent. They are often distinguished by large and small letters, and so a *genotype* is frequently represented as AA, Aa, or aa. Some alleles are dominant over others so that an Aa genotype produces the same characteristic as AA. It seemed logical that dominant alleles would inevitably increase with each generation. However, in 1908 mathematician G. H. Hardy and physician Wilhelm Weinberg showed that populations would reach a balance between dominant and "recessive" genes. The math gave geneticists a base point to compare with real populations. If they found the alleles in a population were not in balance, they knew that another force was operating—natural selection, for example.

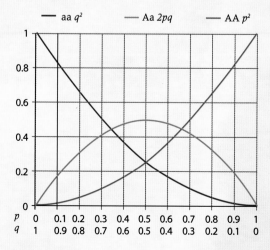

77 Foundations of Mathematics

Bertrand Russell is famous as a philosopher of mathematics but he also tackled the logic of language.

HOW CAN ANY PART OF MATHEMATICS BE PROVEN? THE ANSWER, USUALLY, IS BY SHOWING THAT IT MUST BE TRUE IF THE SIMPLER UNDERLYING MATH IS TRUE ALSO. BUT HOW DO WE PROVE THE SIMPLEST MATH OF ALL?

Starting at the very beginning, how is one to prove that 1+1=2? Between 1910 and 1913 a vast three-volume tome was published on this very subject. It was titled *Principia Mathematica* (The Principles of Mathematics) in a conscious imitation of Isaac Newton's 17th-century masterwork. Its bold aim was to prove the fundamental bases of mathematics by pure logic, and it was authored by celebrated British philosophers Bertrand Russell and Alfred North Whitehead. The first volume concerns itself with the approach to be adopted in the next two, which is through the application of logical *type theory*. In type theory, every mathematical object slots somewhere into an hierarchy of types, each a subset of those above it. This concept is designed to avoid the occurrence of paradoxes, which plague logical approaches to mathematics. The second volume deals with numbers (and does indeed prove that 1+1=2) and the third covers series and measurement. Though the work was a masterpiece, Gödel's theorem was shortly to prove that Russell and Whitehead's attempt—and any other—to prove the entire system of mathematics logically was itself a logical impossibility.

Albert Einstein shares a joke with Arthur Eddington. It was once said that only three people in the world understood the theory of relativity. When it was suggested to Eddington that he was one of the three, he paused before replying. "I'm wondering who the third person is?"

78 General Relativity

BY 1916, ALBERT EINSTEIN'S GENERALIZED THEORY OF RELATIVITY DESCRIBED THE UNIVERSE IN TERMS OF SPACETIME—a combination of four dimensions that required non-Euclidean geometries.

After 20 years in the science spotlight (and a Nobel Prize under his belt), the general theory of relativity cemented Einstein's position as the archetypal scientist. A middle-European accent and eccentric hair was all that was required to evoke scientific genius. The genius of his theory was to not only relate energy to mass but to also bring all the known dimensions into a single cohesive spacetime. Just as Newton's laws of motion were found wanting in describing the Universe on a

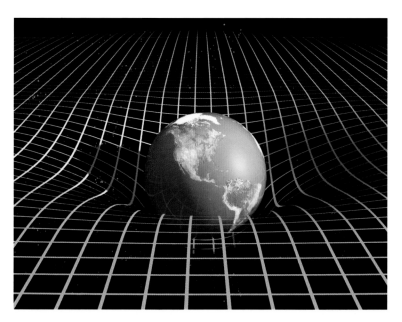

Spacetime can be thought of as a flexible rubber plane, that is bent into "gravity wells" by masses. Larger masses make deeper wells and stronger forces of gravity.

large scale, so was Euclidean geometry. It appeared that straight lines through spacetime were sometimes curves as seen in hyperbolic and elliptical geometry. In 1921, Einstein himself explained, "I place great store by these [non-Euclidean] interpretations of geometry; if I had not known them I would not have been able to develop the theory of relativity."

Bending space and time

Spacetime is curved because of the presence of mass, that is, any object from an atom to a giant star. The curvature is caused by the force of gravity that acts between all masses. As a result the shortest route between two points in spacetime—a straight line—is a curve or geodesic. Light has been observed traveling in straight lines—that fact underlies the whole of optics. But on the grand scales of the Universe with its vast distances and immense masses, the distortions in space time become apparent. The path of light itself curves noticeably—and measurably—as it travels past masses comparable to that of a star. Indeed observing the way our own Sun warped space was used to provide the first experimental evidence of the theory in 1919.

Contraction of space

Non-Euclidean geometry is also required to describe the way spacetime contracts when masses move through it. The theory shows that such contractions are tiny until the speed approaches that of light. At these speeds objects decrease in length (and increase in mass) as they accelerate ever closer to the maximum possible speed.

Objects change shape due to the tidal forces of gravity as well—the gravity pulls harder on the nearest part of the object, causing it to bulge (like the ocean at high tide). In a black hole—a super small, super massive object with the strongest gravity in the Universe— the tidal forces are extreme enough to affect even small objects. A person's feet fall into the black hole faster than the head, causing the body to stretch. This, combined with the contraction of spacetime as the body plummets at great speed toward the black hole, causes a phenomenon called spaghettification, where the body is stretched into a noodle.

THE 1919 "PROOF"

The general theory of relativity predicted that the path of light arriving from stars located behind the Sun—or around its edge at least—would be curved as it passed through the space warped by the star, making it appear out of position. This light was normally drowned out by bright sunshine, but in 1919, Arthur Eddington measured the positions of peripheral stars when they became visible during a solar eclipse. His results supported the theory, catapulting Einstein to world fame. Later analysis showed that Eddington's work was inaccurate, but future experiments confirmed Einstein's theory anyway.

79 The Mathematics of Quantum Physics

FOR 250 YEARS NEWTON'S LAWS OF MOTION AND GRAVITY WERE UNCHALLENGED AS A WAY OF MEANS of predicting the way objects moved through the universe. But at the beginning of the 20th century it became apparent that Newton's laws could not account for everything.

Werner Heisenberg published the uncertainty principle that bears his name in 1927. The uncertainty is a feature of the Universe, not an artefact of our inability to observe. No mater how advanced our technology, we will never develop a detector that can observe the full properties of quantum particles.

One thing that Newton's laws and equations do not adequately describe is the way atoms and subatomic particles behave. Contrary to any commonsense view, matter and energy seem to switch between particles and waves depending on how we are measuring them. We cannot even be sure exactly where a particle is, or its direction of travel. A whole new physics based on the math of probability was required, and from all this uncertainty quantum mechanics was born.

$$\Delta p \, \Delta x \geq \frac{1}{2} h$$

Central Planck

Quantum mechanics tackles the way matter behaves as both waves and particles. Just like any wave, the frequency is inversely proportional to the wavelength, so high frequency matter has a short wavelength. The proportionality constant is the wave's speed, which in the case of the radiation waves tackled by quantum physics is the speed of light, nature's speed limit. The waves released by atoms and molecules were found to be the chief way that energy is transferred between matter. Thanks to Einstein, these were also shown to be a stream of particles called photons. The energy carried by a photon is a specific amount, or quantum, that is derived by another relatively simple bit of math. The energy is proportional to frequency (and inversely proportional to wavelength), resulting in the formula $E = hf$ where h is a value known as Planck's constant, a universal value that was determined by physicist Max Planck in 1900.

Taking a chance

One of the fundamental ideas of quantum mechanics is the uncertainty principle formulated by Werner Heisenberg. No matter how finely tuned our instruments are, because of the wave nature of quanta we cannot simultaneously give accurate values for microscopic particles' position and momentum. On the atomic scale it just is not valid to think of things like electrons as being discrete particles in a defined location. Instead, we have to think of them as smears of probability—we can say where the electron is likely to be but we cannot definitively say where it is.

Quantum mechanics describes the state of any object by a mathematical wave function, from which the possible outcome of any conceivable measurement can be calculated. In general, a quantum system is best thought of as being in a "superposition" of all possible states until an observation is made. Wave functions are solutions of the Schrödinger equation, named for Ernst Schrödinger.

Half-dead cat

Schrödinger is better known for his imaginary cat than for his equation. He proposed a thought experiment to Einstein in which a cat is placed in a box along with a single radioactive atom. When the atom decays it triggers a device that releases a poison, killing the cat. We cannot predict when the atom will decay: all we know is there is a 50-50 chance. There is no way of telling whether the cat is alive or dead. According to quantum theory the atom's wavefunction presents it as both decayed and not-decayed at the same time. Only by opening the box do we *collapse* the atom's wavefunction confirming its distinct state. Until we take a look the math of quantum physics presents the cat as equally dead as it is alive!

A ripple-like pattern is made by the interference of electrons fired through a crystal. Interference is a central characteristic of waves.

WAVE–PARTICLE DUALITY

We think of light and other forms of electromagnetic (EM) radiation as behaving like waves—we talk about wavelengths of light, microwaves, and radio waves. However, we can also measure EM radiation as a stream of particles called photons. EM radiation appears to exhibit what is known as wave–particle duality: it exhibits the behaviors of particles and waves at the same time.

In 1923, Louis de Broglie came up with an extraordinary idea: wave-particle duality was not only a feature of energy, but also of matter. The key point of his idea was that every particle of matter has an associated waveform (not the same as an electromagnetic wave). The faster the particle is traveling, the shorter its associated wavelength. Some physicists mocked de Broglie's ideas, but when experiments were carried out using streams of electrons they were found to behave like waves exactly according to de Broglie's predictions. It was later discovered that this also applied to protons, neutrons, atoms, and molecules. There was no doubt that wave-particle duality is a feature of matter and energy.

80 Gödel's Theorem

FEW THEOREMS OF MODERN MATHEMATICS CAN HAVE ATTRACTED AS MUCH ATTENTION FROM NON-MATHEMATICIANS as that carrying the name of Kurt Gödel. In some discussions, the theorem has been cited as proof of the some bold pronouncements: that minds are "better" than computers; that nothing can be truly proved; that God exists—and God does not exist.

It has been said that Gödel's theorem puts a limit on mathematics in the same way that the Heisenberg Uncertainty Principle limits quantum physics.

What Gödel's theorem proves is not related to any of the above, but it is a startling and influential theory nevertheless, at least within some areas of math and philosophy. Gödel's 1931 paper actually includes two theorems about how mathematical logic is *incomplete*. They both refer to any formal system that can be used to express some simple arithmetic and in which some basic rules of arithmetic can be proven. That covers most of mathematics. Firstly, the theorem says that any consistent system—that is to say one that has no statements that can be proven both true and false—contains statements that cannot be proved or disproved within the system. Secondly no such system can be proved to be consistent within itself.

The key to Gödel's theorem is similar to the ancient paradox of the liar: a person says "This statement is false." If the statement is true, then, as it says, it is false. But if it is false, then that means it must be true. In Gödel's theorem, a similar self-referential statement—but referring to mathematical provability rather than truth—is analyzed.

THEOREM 1

Any theory capable of expressing elementary arithmetic cannot be both consistent and complete. In particular, for any consistent, formal theory that proves certain basic arithmetic truths, there is an arithmetical statement that is true, but not provable in the theory.

THEOREM 2

For any formal effectively generated theory including basic arithmetical truths and also certain truths about formal provability, if it includes a statement of its own consistency then it is inconsistent.

GÖDEL'S TIME MACHINE

Several mathematicians and physicists have used Einstein's general theory of relativity to construct theoretical time machines to travel to the past. By far the earliest of all these devices was proposed by Kurt Gödel. It is fundamental to general relativity that nothing can travel faster than light. One practical reason that would make this impossible is that the mass of a moving object would reach infinity at the speed of light, and so an infinite amount of energy would be required to accelerate an object to such speed. However, by inserting into Einstein's equations a speed in excess of light, the solutions show that the object would begin moving backwards in time.

Gödel found a way round the light-speed barrier: a fast-spinning object distorts the space and time in its vicinity in such a way that the properties of time and space become more similar, until, at high enough spin speeds, journeys back to the starting point in space also allow a return to the starting point in time. But there is a catch: Gödelian journeys like this can take place only in a spinning universe—and there is no reason to believe that we live in one.

81 Turing Machine

THE COMPUTERS THAT HAVE REVOLUTIONIZED OUR CIVILIZATION SO COMPLETELY BEGAN AS A THOUGHT EXPERIMENT IMAGINED BY A BRILLIANT MATHEMATICIAN. His hypothetical machine was never built, nor was it meant to be, but it showed how math could be used, not just to transform data according to a preset system, but also to control the whole process automatically.

The Turing Machine is named for Alan Turing, often elevated to the position of Father of Computing. Turing described his "automatic machine" in 1936, inspired not by a clairvoyant vision of a digital future but as a way of investigating the limits of algorithms in the light of Gödel's incompleteness theorem and to solve Hilbert's tenth problem.

As Turing explained himself in 1948, the machine would have been impossible to make: "...an infinite memory capacity obtained in the form of an infinite tape marked out into squares, on each of which a symbol could be printed. At any moment there is one symbol in the machine; it is called the scanned symbol. The machine can alter the scanned symbol and its behavior is in part determined by that symbol, but the symbols on the tape elsewhere do not affect the behavior of the machine. However, the tape can be moved back and forth through the machine, this being one of the elementary operations of the machine. Any symbol on the tape may therefore eventually have an innings [a British slang term meaning 'to take a turn' or 'be involved']."

The behavior of the machine is governed by an action table of instructions—the algorithms—which relate to what to do with specific scanned symbols in specific circumstances. The world-changing step came when Turing realized this action table could be part of the memory tape, which led to his device evolving into a universal machine capable of performing a computable function that was fed into it. In 1938 Turing met the other "father of computing" John von Neumann. After World War II, von Neumann developed an architecture of Boolean relays that made it possible to put a finite version of the Turing Machine into action: the first digital computers were born.

A Pilot ACE (Automatic Computing Engine) computer built in 1950 in accordance with Turing's design for use at the U.K.'s National Physical Laboratory.

CRACKING ENIGMA

As a leading figure in computing technology it was a given that Alan Turing would be involved in the bid to crack the code used by the Nazis in World War II. The code was ciphered by mechanical Enigma machines, which could generate 159 quintillion different versions of each message. The code was produced by three interchangeable rotors with 1,054,650 combinations. Alan Turing helped to build an electrical system that could produce every one of these combinations within five hours. At 6 p.m. each day a coded weather report was sent out by the German military and it was presumed that it included words like *rain* etc. The code was fed into Turing's device, and the deciphered weather words revealed the key to the rest of the code.

82 Fields Medals

MANY PEOPLE WONDER WHY THERE IS NO NOBEL PRIZE FOR MATHEMATICS—
and there is even a rumor that this "omission" was due to some romantic
rivalry between Alfred Nobel and a mathematician.

In fact, several mathematicians have won Nobel Prizes, such as John Nash (Economics
1994) for his achievements in game theory. However within mathematics itself, the
highest award is the Fields Medal, named after Canadian mathematician John Charles
Fields, who proposed the accolade in 1931 and left some money in his will the year after.

Since 1936, the medal has been awarded at the International Congress of
Mathematicians held every four years. Initially, two medals were given but since 1966
up to four winners have been named each time. Unlike Nobel Prizes, Fields Medal
winners must be under 40 when the prize is awarded, the idea being that the prize
should encourage future progress as well as reward past achievement. The first two
Fields Medal winners were the Finn Lars Valerian Ahlfors and American Jesse Douglas
who both worked on the math of surfaces.

83 Zuse and the Electronic Computer

*Konrad Zuse's Z3 of 1941
was the first "Turing
complete" device, meaning
it simulated a Turing
Machine.*

COMPUTING TECHNOLOGY DID NOT STALL AFTER BABBAGE'S INVENTION. As the
Industrial Age progressed many mechanical devices were built to predict
tides and target the trajectory of artillery shells.

Mechanical computers could be powered by hand—a turn of a handle or wheel—
but powering them by electric motor added to their efficiency. In 1890 Dr. Herman
Hollerith's Tabulator, an electrically powered counting machine, programmed with
punch cards, was used by the U.S. government to process the data for its upcoming
census. Hollerith's Tabulator reduced a ten-year job into one that was finished in six
weeks. Many years later the mighty IBM grew out of Hollerith's company.

A variety of other developments were occurring that many would recognize as part
of computing today, such as the use of Boolean algebra, logic gates, and the binary
notation of 0 and 1 to represent the two-state on/off positions of the computer's
switch-based mechanisms. The earliest example of the modern computer began
construction in 1935, as German inventor Konrad Zuse designed the Z1. This machine

had mechanical switches to store numbers, a keyboard to enter them, and light bulbs to flash answers. Crucially, instructions could also be stored within the machine's memory. However, the biggest advance was programming with binary (digital) notation which sped up the computer compared to the decimal notation used before.

As often happens with major scientific advancement, the early development of the computer was for military applications, and the governments of Germany, the United Kingdom, and the United States worked separately and often in secret. Indeed, the Colossus, built by the British government in 1942, was designed by Alan Turing to break the German Enigma codes.

As technology developed, vacuum tubes replaced gears, and transistors replaced vacuum tubes. The computational powers of computers increased as dramatically as size of the machines decreased. By the 1970s it was possible (though rare) to buy a computer for one's home. Today, computers can be found almost everywhere, from our cellphones to hotel doorknobs.

This room of wires and lights is ENIAC (Electronic Numerical Integrator And Computer) the first general-purpose electronic computer. It was programmed by hand by technicians connecting the switches in specific patterns.

COMPUTER TIMELINE

1642 – Blaise Pascal creates the first calculating machine. It uses eight rotating gears and wheels to add and subtract.

1679 – Gottfried Wilhelm Leibniz establishes the binary system.

1801 – Silk weaver Joseph Marie Jacquard invents a loom that uses punched cards to control weaving patterns.

1822 – Charles Babbage begins work on the Difference Engine, which can calculate mathematical functions (including sines, cosines, and logarithms) up to six decimal places. The machine has hundreds of gears and weighs two tons.

1833 – Charles Babbage designs the Analytical Engine, which has a mill, a calculating unit, an input device (using punched cards), and a printer. There is also a memory store. It is never built.

1859 – England's Registrar's Office commissions a Difference Engine to calculate actuarial tables so as to predict life expectancy—the first use of a computer by a government agency.

1890 – Dr. Herman Hollerith builds the Hollerith Tabulator

1925 – MIT engineer Vannevar Bush and colleagues build an analog calculator that uses electric motors to store values as voltages. Many consider this the first modern computer.

1935 – German inventor Konrad Zuse uses binary notation in his computer designs, increasing speed over the old decimal system.

1936 – Zuse designs the Z1.

1943 – Alan Turing and colleagues design Colossus, a computer tasked to decipher Germany's Enigma codes during World War II.

1950 – Alan Turing writes the first computer program to simulate chess.

1956 – IBM ships the first hard drive. It holds 5 Mb of data and is as large as two refrigerators.

1958 – Jack Kilby of Texas Instruments develops the first model of the integrated circuit. That same year, Seymour Cray designs the first transistor-based supercomputer.

1969 – Intel's M.E. Hoff Jr. designs a processor with 2,250 microtransistors on a chip less than one sixth of an inch long and one eight wide. The Intel 4004 becomes the world's first microcomputer.

1975 – Bill Gates and Paul Allen adapt the computer language BASIC to work with the microcomputer, and found Microsoft to market their achievement.

1977 – the 5 ¼-inch floppy replaces the 8-inch storage medium, while meeting the 8-inch floppy's storage capacity.

1980 – IBM introduces first gigabyte hard drive, as large as a refrigerator and weighing 500 pounds.

1983 – Apple unveils the Lisa, with high-resolution graphics and multitasking. HP releases first computer to use a touch screen.

1983 – First 3.5-inch hard drive released, capable of storing 10 Mb.

1991 – The first 1.8-inch hard drive released, able to store 21 Mb. World Wide Web launched for the public on August 6.

1998 – Google founded.

2012 – Approximately 2 billion people have access to the Internet.

84 Game Theory

Game theory was just one of the many world-changing achievements of John von Neumann. The Hungarian-born mathematician also designed early digital computers and worked on the Manhattan project.

IT WAS A LITTLE IRONIC WHEN, AFTER HALF A CENTURY OF MATHEMATICIANS TRYING TO CONVERT MATH INTO A SET OF NON-NUMERICAL IDEAS, as the computer age dawned, new technology produced a demand for a system that could render ideas as a series of numbers. One result was game theory, one of the most crucial pieces of applied math ever conceived.

Math and games have a long association. The field of probability grew out of a desire to gamble most effectively during dice games, while the Bridges of Königsberg problem, where strolling vacationers tried to trace a specific route through the Prussian city, gave rise to graph theory and topology. Breakthroughs have frequently been made in idle moments, while at the same time mathematicians from Diophantus to Lewis Carroll have left math puzzles in their work.

The mathematicians and economists of the Rand Corporation discuss how to allocate the resources of the Strategic Air Command, which maintains the American nuclear arsenal. Game theory was used to ensure that nuclear weapons were deployed to their fullest potential at the height of the Cold War in the late 1950s and 1960s.

What are the chances?

The actual math of games falls into groups—those that involve chance, and those that do not. The latter group can be analyzed to find the winning strategy, which if correctly followed will guarantee victory. A variant of this is tic tac toe (or noughts and crosses) where both players will ensure a tie if they follow optimal strategy. A game of chance involves calculating the probabilities of winning or losing events. In 1928, American scientist John von Neumann founded the field of game theory, in which probability and strategy were applied to real-world situations.

CUBAN MISSILE CRISIS

Game theory had resulted in equilibrium in the Cold War nuclear arsenals: the mutually assured destruction strategy (MAD) ruled out war because an attacker would come off as badly as the enemy. However, when Soviet missiles were stationed in Cuba in 1962, the game was changed—America could be hit within minutes leaving no time to retaliate. For 13 days the world stood on the brink of nuclear war until both sides agreed to pull back.

Zero-sums and pay-offs

The basis of game theory is the zero-sum game, where a benefit to one party results in a loss of equal magnitude to the other side. Gains and loses always add up to zero so there is never a need to cooperate with the opponent. The predicted outcome of a player's actions and the actions of their opponents are displayed in a payoff matrix, like the one on the right. This shows the effects of a policy pledge during a mayoral election between the incumbent Martha and opponent Ruth. The matrix shows that the best thing for Martha to do is to campaign for a stadium on the east side. Ruth's best strategy is to pledge to not build the stadium, and so win more votes.

These strategies are independent of each other and so the game is described as "determined." In another scenario, Ruth's best option might change based on what Martha does, in which case a minimax strategy is used, which minimizes the maximum loss based on the probability of the strategy of the other side. When games are not zero sum, and both sides can win or lose, the possibility of cooperation appears. The question is how sure can you be your fellow player will stick to the deal?

This matrix shows how many votes Martha will get depending on where she agrees to build the new city stadium. Ruth gets the rest in the zero-sum game.

		Ruth Ival		
		West	East	No Stadium
Martha Ayor	West	55%	45%	35%
	East	60%	65%	45%
	No Stadium	45%	50%	40%

85 Information Theory

AS MORE COMPLEX PROGRAMS WERE WRITTEN AND STREAMS OF DATA **PASSED FROM ONE COMPUTER** to another, mathematicians found they had something new to measure—information itself.

In the context of computing the word bit is a contraction of binary digit, and stands for a single 0 or 1 (the only two possibilities). It was coined by Claude Shannon in 1948. In 1956 Werner Buchholz coined byte for a set of eight bits. The code for a letter or other keyboard character was 1 byte and the smallest amount, or bite, of data a computer could process at a time—byte was respelled to avoid confusion with bit. In the 1970s the nibble was introduced, named because it had 4 bits and was smaller than a byte.

Claude Shannon is the forgotten father of information theory. One of his contributions was the parity code as a mathematical means of detecting corrupted data. The code was sent as bits attached to the end of the original information. In the example on the left, codes are added to nibbles by adding the bits 1, 2, and 3, then bits 1, 2, and 4, and finally 2, 3, and 4. If the results of the sums are even then a 0 parity bit is added. Odd results give a parity bit of 1.

Original message	Sent message
0000	0000000
0001	0001011
0010	0010111
0100	0100101
1000	1000110
1100	1100011
1010	1010001
1001	1001101
0110	0110010
0101	0101110
0011	0011100
1110	1110100
1101	1101000
1011	1011010
0111	0111001
1111	1111111

86 Geodesics

A GEODESIC IS TO A SPHERE WHAT A STRAIGHT LINE IS TO A FLAT SURFACE. GEODESIC GEOMETRY HAS ITS USES FROM THE "GREAT CIRCLES" OF longitude that girdle the Earth, to the gravitational curves of spacetime. But in 1949, it was also put to use in one of the most iconic construction designs of the 20th century: the geodesic dome.

One thing a straight line and a geodesic have in common is that they are both the shortest ways between two points. This is why a flight path displayed on a flat map of the world is often seen to curve eccentrically. It is follows a geodesic, otherwise named when it comes to Earth a great circle or meridian. If the planar map was wrapped around a sphere, the route would look a lot more direct.

However, straight lines and geodesics have more differences than similarities: a straight line is infinite in length, while a geodesic is closed, always circling back on itself. Straight lines can also be parallel, while geodesics never are.

The lines of longitude on a globe are geodesics, but the lines of latitude are not (except for the equator).

Efficient space

The geodesic dome was patented by Richard Buckminster Fuller. He was interested in efficiency and designed prefabricated sections that spread the weight of the entire structure, meaning the domes could enclose very large spaces (the largest is 216 meters across). Buckminster Fuller was inspired by crystal structures, and two years after his death in 1983, a new type of carbon formed as geodesic spheres was discovered and named buckminsterfullerene in his honor.

PLANETARY PASSENGERS

In his 1969 book *Operating Manual for Spaceship Earth,* Richard Buckminster Fuller likened planet Earth to a spacecraft and humans to its passengers. In a view perhaps outmoded today he described how energy from the Sun fueled Earth, flowing through the biosphere and harnessed by civilization to make the things we need before being radiated into deep space. Fuller recognized that the energy available to Earth was constantly decreasing. However, he suggested that human knowledge was constantly increasing. Each increase in knowledge, he argued, allows people to harness the resources available to increase their collective wealth and standard of living. If even some of the predictions about the technological value of the fullerenes prove true perhaps they will form a new chapter in the planet's operating manual.

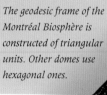

The geodesic frame of the Montréal Biosphère is constructed of triangular units. Other domes use hexagonal ones.

87 Chaos Theory

BY THE MID **19**TH CENTURY, THERE WERE MANY WHO BELIEVED THAT MOST SCIENTIFIC QUESTIONS ABOUT THE **U**NIVERSE WERE EITHER ANSWERED OR CLOSE TO BEING ANSWERED. Then math spoilt this comforting view—natural phenomena were actually chaotic.

The Universe was once thought to be like a well-tuned watch, set into motion by a benevolent creator and left alone to tick along under the principles of physics and mathematics. This view of the universe was Newtonian, named for Sir Isaac Newton and his seventeenth-century laws of motion and gravity. However, there were unanswered questions, and one of these was the three-body problem, where Newton tried to explain the motion of the Moon within the gravitational influence of the Earth and the Sun.

By the late 1880s the French genius Henri Poincaré took the first steps towards what would become chaos theory. He noted that tiny changes in the velocity or position of three gravitationally interacting bodies would be amplified over time to give completely different behaviors. Another example is the driven double pendulum, that is, a pendulum supported from a vibrating point. Its behavior depends on both the swing of the pendulum and the frequency of the vibration and may change markedly when one or the other frequency is changed.

Chaos theory is often described using the Butterfly Effect, a term coined by Edward Lorenz in the 1960s. It refers to how a tiny change, such as the air moved by the beat of a butterfly's wings, can lead to a large result, such as the high winds of a storm system.

Chaos ensues

Work on the problem came up against an obstacle: a lack of computational power needed to perform the calculations. So the field became dormant for many decades, until American meteorologist Edward Lorenz came across it while modeling weather systems in 1961 using a primitive digital computer. He found that the models produced completely different results with every initial condition—even conditions that appeared in the processing data from other iterations of the same model. The differences were due to tiny inaccuracies between the number processed by the computer and the ones it printed out, which were rounded up. Such tiny differences were thought to be inconsequential, but Lorenz showed that in chaotic systems small initial changes produce entirely different outcomes.

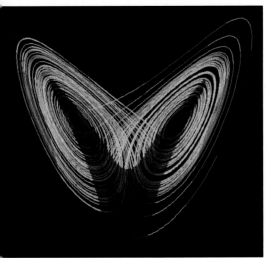

The Lorenz Attractor is a plot of the chaotic solutions for a set of differential equations studied by Edward Lorenz to predict the weather. It is an early example of a fractal.

POINCARÉ'S *N*-BODY PROBLEM

In the 1880s, Henri Poincaré came across chaos in the three-body problem (and, later, the *n*-body problem), and proved that orbits could be non-periodic—without a fixed time period for each revolution. However, despite the erratic motion, the body does not move away from or approach a fixed point. What sort of mathematics could describe such chaotic movements? Poincaré was not able to fully solve the *n*-body problem, but he made such progress that he was awarded a prize for his work by the King of Sweden in 1887.

88 String Theory

THROUGHOUT THE 20TH CENTURY MATH HAD SAVED PHYSICS FROM EATING ITSELF WITH THE THEORIES OF THE VERY SMALL AND OF THE VERY LARGE. What could not be observed or made to conform with human perceptions were rendered as purely mathematical models. By the 1960s math offered a means to tie physics back together.

It is hard to imagine string theory beyond wobbling loops and lines. The addition of compact dimensions transforms the linear strings into multidimensional surfaces or manifolds which interact with each other over time "creating what is called a worldsheet"

Albert Einstein had helped to bring the quantum world into focus with his postulation of photons as energy carriers for light and other radiation—and then his theories of relativity tackled the Universe on the very large scale. All this was achieved in about 10 years of inspiration (and perspiration) and Einstein spent the next 40 of his life trying to bring both theories together in a theory of everything. This meant coming up with a way of linking fundamental forces of nature such as gravity and electromagnetism.

Points mean strings

The search for the theory of everything continues to this day, but what is generally accepted as the front runner appeared in the late 1960s, with the help of far-reaching fields of math such as group theory and topology. The result was string theory which represents subatomic particles not as zero-dimensional points but as one-dimensional lines, or strings, A string's oscillation, once captured by mathematics, defines its properties, such as spin or charge. Much of the vibration takes place within so-called compact dimensions. These exist only on the quantum scale but allow the strings to move in several directions all at once—the latest theory requires 11 dimensions.

The initial forms of this theory, such as the dual resonance model of 1969, focused on the bosons, particles such as photons and others that were responsible for mediating fundamental forces. By the 1990s the strings had been extended into superstrings which formed a connection between the bosons and the fermions—electrons, quarks, and other particles that give matter its mass. The connection between the massless bosons and massive fermions is called supersymmetry and this is one area that is now under investigation thanks to the 2012 discovery of the much-vaunted Higgs boson.

BIG MEETS SMALL

String theory and its descendants have had many critics to say it is a philosophy rather than scientific theory since there is nothing to experiment on. One place that the theory is most readily tested is in a black hole, where the theories of big and small scales collide. A black hole is a star collapsed into essentially a single point, forming a mind-boggling tangle of strings on the quantum scale. However it is also massive enough to have the strongest gravity of any object in space. Quantum uncertainty tells us that within the shortest possible time period (10^{-43} seconds) virtual particles, or strings, of matter and antimatter exist everywhere, ceaselessly forming and then annihilating each other. The opposing virtual particles on the edge of a black hole become separated by the gravity in the instant of their formation. This results in one string being released from the hole, providing vital clues as to what the exotic, primeval matter inside is like.

89 Catastrophe Theory

WHILE NOT DIRECTLY RELATED TO CHAOS THEORY, CATASTROPHE THEORY EXAMINES HOW SMALL CHANGES IN CIRCUMSTANCES CAN PRODUCE SUDDEN AND LARGE SHIFTS IN BEHAVIOR. A good analogy is how a mountainside can be stable for many millennia and then a single rain shower or earth tremor makes it collapse in a landslide.

The mathematical way of explaining this catastrophe and other events like it—the calving of icebergs or the crash of a stock market— is to represent the system as an equation. This equation is in equilibrium over a long period of time no matter what the variables within it are doing. What element in that equation can cause a bifurcation, a sudden shift that results in a catastrophic event?

Active variables

French mathematician René Thom started work on catastrophe theory in the 1960s, and the work was popularized by the British mathematician Christopher Zeeman in the 1970s. These mathematicians classified various types of catastrophes based on the number of active variables at play. Some of this mathematics has been applied to the capsizing of ships, bridge collapses, and panic buying.

Another type of catastrophe has been called the "tipping point," where one or more active variables build up to a critical level, after which their impact on the system multiplies rapidly and becomes impossible to contain. Such a phenomenon is one theory under examination in current models of human-made climate change.

Icebergs are chunks of the ice sheet that break off as the weather warms up in the polar summer. Catastrophe theory is being used to examine how changes in the rate icebergs are "calved" and how long they stay afloat before collapsing can be indicative of climate changes.

To some the most catastrophic event of all is a crash in the stockmarket, where confidence in investments takes a severe downward adjustment after a minor drop in a few prices.

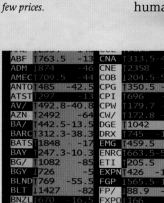
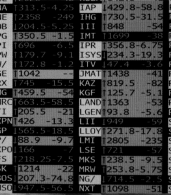

90 Four-Color Theorem

A COLOR MAP IS A CLEAR MAP, WITH EVERY REGION GIVEN A DISTINCT COLOR. **T**HE DEVELOPMENT OF PRINTING TECHNOLOGY and the rise of the nation state in the 19th century resulted in some very colorful maps indeed. To simplify the process—and keep costs down—cartographers asked the question how many colors does a map really need?

In Mark Twain's *Tom Sawyer Abroad* (1894), Tom and Huck Finn are arguing about their progress on a balloon flight heading east from St Louis. Tom estimates from the wind speed that they have reached Indiana, but Huck contends that they have been given false information: "We're right over Illinois yet, you can see for yourself that Indiana ain't in sight... Illinois is green, Indiana is pink. You show me any pink down here if you can." When Tom protests, Huck insists, "...I've seen it on a map, it is pink."

To all but Huck Finn, a colored map generally reduces confusion, and with a bit of intuition followed by trial and error, cartographers found that the maximum number of colors needed is four. Any map with clear boundaries between one region and the next, be it of the world's nations, city districts, or the floor plan of the most opulent fantasy palace, requires just four colors to ensure that no two neighboring sectors are colored alike. In 1852, mathematician Francis Guthrie wondered why this was so, and produced maps as graphs, with regions as faces, frontiers as edges, and vertices where three or more regions met. However, no one could find the proof—despite map makers never needing a fifth color. The story finally ended in 1976 back in the still-green Illinois, when Kenneth Apel and Wolfgang Haken used the university's mainframe to show that an infinite number of countries would only ever require four colors.

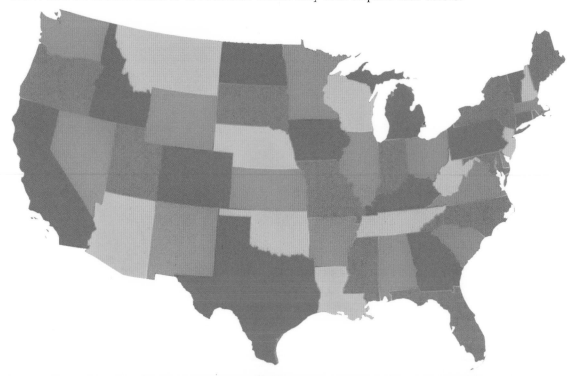

The four-color theorem at work on the lower 48 U.S. states. Putting the map on a globe would not require any changes but if Earth was a donut (inedible, obviously), we would need seven colors.

91 Public Key Encryption

As THE POWER OF COMPUTERS INCREASED, SO DID THEIR IMPORTANCE TO A COUNTRY'S GOVERNMENT AND MILITARY—and they became crucial tools in economic transactions. The need to keep data secure became clear and mathematics came to the rescue.

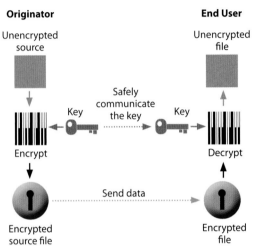

Originator
Unencrypted source
↓
Encrypt
↓
Encrypted source file

Key ← → Safely communicate the key → Key

Send data

End User
Unencrypted file
↑
Decrypt
↑
Encrypted file

Since it is impossible to prevent eavesdroppers picking up secret communications or collecting sensitive information, the only method to protect data was through the use of encryption. But then again computers had already been used to defeat encrypted data, such as the Germans' Enigma codes in the Second World War.

One-way math

Early work on mathematical data encryption dates back to 1874, when mathematician William Stanley Jevons developed operations that were relatively easy in one direction, but reversing the process was much more difficult and disproportionately time-consuming. A hundred years later, cryptographers Whitfield Diffie and Martin Hellman took this into account when they created the Diffie-Hellman key exchange. An analogy is that two people wish to send secret messages to each other. They have a lockable box, their own padlocks, and their own keys to those padlocks. The first person asks the second person for their open padlock. Placing the message in the box, the first person locks the box with the second person's padlock and sends it to the second person. The second person uses his or her key to open the padlock and access the secret message.

Against the clock

Public key encryption uses extremely large numbers on the understanding that there is no efficient means of factoring large numbers. A third party might eventually succeed in decrypting the message but it would likely take so long that the information would no longer be useful. Today, the best online security uses 128 and 256-bit encryption. In 128-bit encryption, someone hoping to guess the correct number to decipher the secret message would have to work through 2^{128} permutations. A brute force attack (working through every possible permutation) could take 149 trillion years. A 256-bit key would take that amount of time, squared!

DIFFIE-HELLMAN ALGORITHM

In 1976, Americans Whitfield Diffie and Martin Hellman published a method for two people to send private coded messages without having to exchange any key to the code in advance. The system uses modular arithmetic, as well as the properties of prime numbers:

1. Andy picks a number that he keeps secret. We call this number A1.

2. Zelda picks another number that she, too, keeps secret. This number is Z1.

3. Next, both Andy and Zelda apply a function of the type $f(x) = a^x$ mod. p, with their respective numbers being x, while p is a prime number known by both and a is the publicly visible information. From this operation Andy obtains a new number, A2, which is the remainder of the a^x divided by p. He sends that number to Zelda. Performing the same operation, Zelda obtains a new number, Z2, which she sends to Andy.

4. Andy solves $Z2^{A1}$ mod. p and produces another new number, CA.

5. Zelda solves the equation $A2^{Z1}$ mod. p to produce a new number, CZ. Amazingly CA and CZ are the same and this is the key that both use to decipher the message.

92 Fractals

TRADITIONAL GEOMETRY DEALS WITH STRAIGHT LINES AND PERFECT CIRCLES BUT IN THE REAL WORLD NONE OF THESE EXIST. Clouds, trees, and rocks are fragmented, jagged, and fractional. Nature is rough and lumpy. Describing such real-world systems has historically defeated mathematicians. The devil, it appears, is in the detail.

As the name suggests, fractal mathematics is an attempt to come to terms with the fractured roughness of everyday life. The Belgian mathematician Benoit Mandelbrot coined the term in 1975, to encompass his explorations into a new geometry that is "equally rough at all scales." His starting point was a simple problem involving the measurement of coastline.

On a large-scale map, the coastline of an island is very uncomplicated. Because the level of detail is low, the edges can be represented with straight lines. To get more detail you need to zoom in. A measurement at this new scale requires a smaller ruler and, by reducing the size of the measuring stick, one can get an increasingly accurate estimate of the length of the coastline. There is, however, no limit to this process and as one continues zooming in, the calculation becomes ever larger, and it is impossible to find a final figure before the edge of the land becomes indistinguishable from the water.

Getting lost in the detail

These so-called "pathological curves" give the shapes you see in nature. The body is covered in fractals—there is detail and complexity at all scales of magnification. The lungs, for example, fill space incredibly efficiently. With their innumerable folds and crenulations, the volume of the lungs is very small but the surface area is enormous.

Plants, too, grow using simple rules yet reach great complexity. The convoluted surface of the Romanesco broccoli makes measuring its area difficult. The closer you look at the repeating patterns of florets and whorls, the more detail you see. The method of growth by addition of repeated units on several scales seen in the Romanesco broccoli gives rise to a phenomenon called "self-similarity."

In 1904, the Swedish mathematician Helge von Koch discovered a construction that would imitate self-similarity. His "Koch curve" takes a "seed shape" and then replaces the middle third of each side with a scaled-down version of the original shape. When the seed is an equilateral triangle, the next stage is a six-sided "Star of David" and very soon, a snowflake emerges.

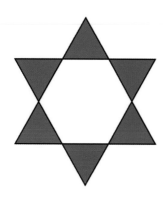

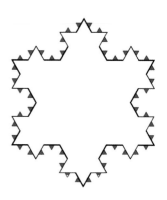

Fractals often reflect the shapes of natural features. The details of a Mandelbrot set (right) resemble the spidery side channels of this immense reservoir (left). Fractal geometry is used in computer-generated imagery (CGI) to make landscapes and ocean waves look realistic.

Complexity encapsulated

Each part of a self-similar shape, no matter what scale you look at it, resembles the whole thing. Investigating such shapes was limited to hand drawings until the advent of computers. The Mandelbrot set, perhaps the most widely reproduced image in mathematics, first appeared in the 1980s when the computing power that made it possible became available. For all its complexity, the math behind the M-set is not complicated, and involves only adding and multiplying numbers: $z = z^2 + c$. The key is iteration—simple rules repeated without end. The output becomes the next input and so forth. When Mandelbrot was investigating this phenomenon (that he called his "big bear") in 1980, he found that for certain starting values of z the outputs would continue to grow forever, while for others they shrunk to zero. The M-set then is the boundary limit between these two classes of number. Outside the lines are free z-values bound for infinity; inside are prisoners, destined for extinction. Zooming in on any boundary resets the scale of the numbers, but the patterns are always the same.

Every object has a fractal dimension, which is a kind of statistical "roughness measure." The Romanesco broccoli has a fractal dimension of around 2.8, a coastline is 1.28, and human lungs are about 2.97. Formulas for trees, clouds, and mountain ranges can be generated entirely artificially based on a measure of their complexity. Scientists also look for fractal patterns in data sets that might otherwise appear chaotic.

THE PEANO CURVE

In 1890, Italian Guiseppe Peano discovered a one-dimensional curve that could extend across a two-dimensional plane in perpetuity. Using a construction similar to, but pre-figuring, the Koch curve, Peano started by mapping a unit interval onto a unit square and repeated the process with smaller intervals. He produced an infinitely long line that does not spread beyond the square. Peano never actually drew the curve, preferring to prove this counter-intuitive property mathematically.

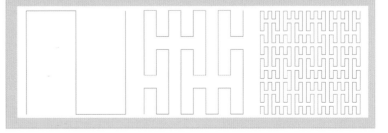

93 The Fourth Dimension and Beyond

WHAT IS THE FOURTH DIMENSION? STUDENTS OF EINSTEIN MIGHT VENTURE THAT TIME IS NUMBER FOUR AFTER THE THREE SPATIAL DIMENSIONS—LENGTH, WIDTH, AND DEPTH. However, mathematicians are seldom bounded by the laws of physics, and so conjured shapes with ever more spatial dimensions. That fact that we cannot see them is neither here nor there—quite literally.

In 1884, the English teacher Edwin Abbott Abbott published *Flatland: A Romance of Many Dimensions*. This book was not just a biting satire of the inequalities of Victorian Britain, but it also elucidated the links between perception and dimensions, and did much to establish the idea of the fourth dimension as "the other world" or "higher plane" that can be traced through science fiction to this day.

Flatlanders were, as you might expect, entirely flat and occupied a Cartesian plane. The story involves a Square taking a journey to Lineland, where everyone is a one-dimensional line or a zero-dimensional, lengthless dot. He is then visited by a sphere from a three-dimensional world, who appears to him as a circle that grows and shrinks as it passes through Flatland. Similarly, a human is a three-dimensional creature. We can perceive the distance in space and also experience material within those three dimensions changing—that is what we know as time. If we were to see a four-dimensional object moving through space, all we could perceive would be its three dimensions changing over time.

By the 1980s, geometry had gone extradimensional. If a line stops and starts along the *x* axis, and a square is on *x* and *y*, then a cube has coordinates on the *x*, *y*, and *z*. It is simple enough to add a *w* axis for the fourth dimension, to get a 4-cube, or hypercube. An axis is added for every higher dimension, or algebra represents *n*-polytopes, any flatsided, smooth shape with *n* dimensions.

The Grande Arche built in 1989 at La Défense in Paris is a tesseract, a representation of a hypercube, or four-dimensional cube. It has twice the number of corners (vertices) as a cube and four times the faces and edges.

LIVING IN TWO DIMENSIONS: THE CHARACTERS OF ABBOTT'S *FLATLAND*

Edwin Abbott Abbott wrote his *Flatland* story under a pseudonym, A. Square. This was probably an arch joke about his repeated name, but is also the name of the chief protagonist. Flatland society was populated by geometric figures, the shape of which indicated their position in society. The working class were triangles, the more acute the angle of their apex the more menial their roles: soldiers and servants were pointed slivers, while equilateral triangles were the administrators and shop keepers. A professional like our narrator was square; a gentleman of leisure pentagon, and a hexagon nobleman has one more side. The highest echelon were the priests, where the number of sides has become so high, that they have become a curved circle. Abbot's decision to make all women a one-dimensional line has drawn criticism since, but he used it as a comment on the social restrictions placed on 19th-century women, not their apparent simplicity compared to men. The cover of *Flatland* had a picture of Square's house—note how the sons add their mother's line to move up a social class.

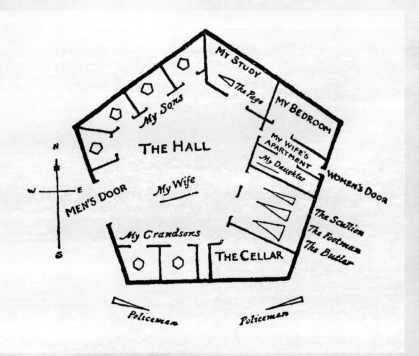

94 Classification of all Simple Finite Groups

THESE ARE THE BUILDING BLOCKS OF MATH. JUST AS EVERYTHING IN THE UNIVERSE IS CONSTRUCTED FROM CHEMICAL ELEMENTS, so every finite group is built up from a limited number of simple groups.

Given the power of groups in both mathematics and science, the classification of all simple finite groups became a major 20th-century math project, carried out by more than 100 mathematicians and resulting in more than 500 papers. It was completed in 1985. It turned out that there are just 18 kinds of simple finite group, together with 26 unique groups. The most celebrated—and largest—of the unique groups is known as the Monster. Mathematicians are not prone to hyperbole: the group is truly monstrous, being composed of 808,017,424,794,512,875,886,459,904,961,710,757,005,754,368,00 0,000,000 elements measured in 196,883 dimensions. While the Monster was isolated simply to insure that the classification of all finite simple groups was complete, its form was hauntingly familiar to mathematicians working in what was believed to be the entirely separate area of modular forms. Analysis revealed not only deep linkages between these two areas but threw significant new light on quantum theory, too.

95 Self-Organized Criticality

THE 1987 PAPER *SELF-ORGANIZED CRITICALITY: AN EXPLANATION OF THE 1/F NOISE* GOES TO THE HEART OF ONE OF THE MOST mystifying aspects of our Universe: we are surrounded by complexity of the most baffling kind, and yet the equations that explain them are—relatively—simple.

The theory is potentially applicable to a vast range of processes, from wars to earthquakes and to explain why the sand dunes in a desert have the same shape, but different scale, as the ripple-like ridges on a sandy beach.

Per Bak, Chao Tang, and Kurt Wiesenfeld's paper has become one of the most cited of all on a mathematical subjects, and with good reason. We know, for example, all the laws that affect the weather and we have plentiful observations of meteorological conditions as well as powerful computers, and yet our weather forecasts 50 days in the future are hopeless. By studying cellular automata (simple self-governing systems), Bak, Tang, and Wiesenfeld were able to show that complex behaviors occur as a matter of course, independent of outside influences and for a wide range of initial states.

96 Fermat's Last Theorem

FEW MATH PROBLEMS HAVE CAPTURED THE PUBLIC IMAGINATION AS THIS ONE. ITS EVENTUAL PROOF MADE FRONT-PAGE NEWS, spawned best-selling books, and its British conqueror was knighted by the Queen. But at first glance it is intriguingly simple: $x^n + y^n \neq z^n$.

Fermat was the ultimate mathematical puzzler.

The story of Fermat's last theorem begins in the 1630s and spans the next 360 years. The reason for its long run is because the math behind it is fiendishly complex. Andrew Wiles, the man who proved the theorem in 1994, described the theorem as a large house, a mansion plunged into pitch darkness, with the answer somewhere within: "One enters the first room of a mansion and it's dark. Completely dark. One stumbles around bumping into the furniture, but gradually you learn where each piece of furniture is. Finally, after six months or so, you find the light switch, you turn it on, and suddenly it's all illuminated. You can see exactly where you were. Then you move into the next room and spend another six months in the darkness..." Wiles's fame is based on a note scribbled in the margin of page 85 of a copy of *Arithmetica*, the seminal work by Diophantus. The book belonged to Pierre de Fermat, a lawyer from southern France. It said "If n is an integer greater than 2, then there are no integers x, y and z other than 0, that fulfil the equation: $x^n + y^n = z^n$." Easy enough to say but it true? The note went on "I have discovered a truly marvellous proof, which this margin is too narrow to contain."

Puzzles then proofs

Fermat made a habit of sending such mathematical *bon mots* to his friends in Paris—often disseminated by the mathematical monk Marin Mersenne. As an amateur mathematician Fermat did not need to offer proofs but saw these missives as fun challenges. He often said he had the answer but frequently failed to make it available if no one else could find a solution.

Andrew Wiles enjoying his lecture on his proof of Fermat's Last Theorem in 1993. However a last-minute inconsistency meant he had to work for a further year to get his proof accepted.

There is no record of this final puzzle being sent out for scrutiny, and it was only revealed when Samuel de Fermat, the great man's son, was cataloging his papers for publication after his death in 1665. Try as he might, Samuel could not find the promised proof to this "last theorem" among any other documents. Many have suggested that if Fermat had a proof at all it would have been incomplete judging by the centuries of turmoil that followed as professional mathematicians tried to find it. In 1995 Andrew Wiles's paper *Modular Elliptic Curves and Fermat's Last Theorem* was accepted as a conclusive proof. It had taken Wiles eight years of intense work and more than 100 pages to explain that which Fermat could not scribble in the margin.

97 Proof by Computer

THERE IS SOMETHING SPECIAL ABOUT DISCOVERY; IT REQUIRES SOME IMAGINATION, OFTEN ARRIVING FROM BEYOND REASON, STARTING ON A HUNCH. There is something deeply human about it. Or there was until 1996 when the first math problem was solved by a computer program.

The first problem to be proven by a machine was Robbins' conjecture which was concerned with the interchangeability of two axioms of Boolean algebra. The problem was posed by Herbert Robbins in 1933 and several leading figures in the field were unable to close the debate. Then in the 1990s, William McCune, a scientist at the U.S.'s Argonne National Laboratory, proved that Robbin's axiom did produce Boolean algebra, or rather his program EQP did. EQP stands for "equational prover" and was the first of several automated theorem provers. These programs can tackle first-order problems, which are based on axioms, rather than derivations of such. The computer provides the brute force to solve these problems, however a future milestone might be the first mathematical conjecture named after a computer.

98 Millennium Problems

AT A 1900 CONFERENCE IN PARIS, DAVID HILBERT HAD SET OUT 23 MATH PROBLEMS FOR THE COMING CENTURY. A century later, mathematicians again gathered in the same city to hear what challenges were to occupy them in the 21st century.

The logo of the Clay Mathematics Institute is the "Figureight Knot" representing the "orbifold X given as a quotient of three-dimensional hyperbolic space."

The Millennium Prize Problems, as they were called, were presented by the Clay Mathematics Institute (CMI), set up the year before in Massachusetts by the wealthy patron Landon T. Clay. Seven problems were listed, each one with a $1 million prize. As Hilbert did before them, the advisory committee to the CMI—containing such luminaries as Andrew Wiles and Arthur Jaffe—chose problems that would bear the most constructive fruit: P versus NP problem, Hodge conjecture, Poincaré conjecture, Riemann hypothesis (the only hangover from Hlibert's list), Yang–Mills existence and mass gap, Navier–Stokes existence and smoothness, and Birch and Swinnerton-Dyer conjecture. As yet only one has been solved, the Poincaré conjecture.

99 Poincaré Conjecture

IN 1904, THE FRENCH MATHEMATICIAN HENRI POINCARÉ POSED ONE OF THE WORLD'S great mathematical conundrums. It took 98 years for him to be proved correct.

A leading light of the mathematical community and father of chaos theory, Poincaré followed a strict regime of work—two hours in the morning and two hours in the evening—allowing his subconscious to take the strain of heavy conceptual work in the meantime. Perhaps while musing in his rest period, Poincaré speculated that the sphere is the simplest possible shape in any dimension

The Poincaré conjecture was proved correct by Russian Grigori Perelman in 2002. He has since refused all awards for this momentous achievement.

Simply connected

As with any mathematical problem the terms need defining, and Poincaré used a two-dimensional loop to demonstrate the simplicity of his three-dimensional shape. A space is "simply connected," he said, if every loop upon it can be contracted to a point. Imagine a noose around a slippery ball. No matter how the cord is slung, the ball will escape as the noose tightens. Not so with a donut, as the noose closes around its central hole it will snag and pull tight. A donut is not simply connected. Poincaré wondered

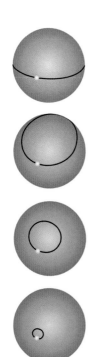

A diagram of the simple connectivity of a sphere, with a circumference contracting to a point.

whether this held true in more than three dimensions. Was the sphere always the simplest shape, or were other shapes simply connected in those realms?

Seeing them all

The unproven conjecture had bested leading minds for almost a century, and was the third of the Clay Millennium Problems. Two years later, a Russian-born mathematician had the answer. In three papers published in 2002 and 2003, Grigori Perelman solved this topological problem using a technique he called "Ricci flow with surgery." Several teams of mathematicians have since verified the results and Perelman was awarded the $1 million Clay Institute prize and the 2006 Fields Medal for his proof. He refused both, seemingly unimpressed with his achievement. As he put it, "A simply connected closed 3-manifold sounds complicated but trust me, once you've seen one you've seen them all." While everyday explanations of the ramifications of the Poincaré conjecture are hard to come by, the math may help understand the shape of the universe as it has expanded since the Big Bang.

PERELMAN TURNS DOWN THE PRIZE

Uninterested in money or fame, Grigori Perelman turned down his 2006 Fields Medal, the mathematical equivalent of the Nobel Prize, and has not collected the one million dollar prize offered by the Clay Mathematics Institute for proving the Poincaré conjecture. He lives with his mother in St Petersburg and will not to speak to journalists, as he claims they are only interested in why he refused the million and whether he cuts his nails. His most famous quote is *"I don't want to be on display like an animal in a zoo."* Perelmen considers his contribution to be no greater than that of U.S. mathematician Richard Hamilton, who introduced the Ricci flow technique that he built his proof upon. Perelman's humility and purity of purpose has won him worldwide admiration, although many believe that his refusal to claim the prize money is the moment when genius and wisdom parted ways.

100 The Search for Mersenne Primes

A PRIME NUMBER PROBLEM SET BY A **F**RENCH MONK IN THE **17**TH CENTURY is being solved today by a Internet-linked math community.

$$M_p = 2^p - 1$$

The French monk Marin Mersenne is remembered for his scientific salon in Paris where he presented the works of Descartes, Fermat, and Galileo and which became the model for Europe's great academies of science. However, Mersenne also has a subset of prime numbers named after him. A Mersenne prime (p) is one that when raised as the power of 2 and 1 is subtracted, the result is also a prime number (M_p). Finding these numbers is not easy and in 1996, the GIMPS (Great Internet Mersenne Prime Search) began. This invited computer users to donate their PC's processing power to help with the math. GIMPS is performing 68 trillion calculations per second. The last new Mersenne prime was added in 2009. There are currently just 47 known; the largest has almost 13 million digits.

The only way to find a Mersenne prime is by trial and error, plugging primes into the formula. The search for M_{48} continues at mersenne.org.

Mathematics: a guide

MATHEMATICS IS ALL ABOUT BEING PRECISE. EVEN THE EXPRESSION OF IMPRECISION IS DONE IN A VERY PRECISE WAY. That is the beauty of numbers: they can only mean one thing, unlike words. Much of mathematical language can be confusing because it has a precise mathematical meaning but uses terms drawn from everyday language—so it is easy to get lost. To help you find your way again, here we attempt to bust some of that jargon.

Quantities

WHEN IS A NUMBER NOT JUST A NUMBER? WHEN IT IS ONE OF THESE TERMS. This section looks at the words used to describe numbers or groups of numbers with specific forms or origins, and other associated terms.

Binary: Relating to a counting system that uses just two numbers, 0 and 1. Binary is the language of computing, each number or digit is a "bit" of data, eight of them add up to a "byte".

Cardinal number: Numbers that refer to quantity.

Composite: A number that can be divided by a number other than itself. Non-composite numbers are primes.

Decimal: Relating to the base ten counting system; decimal fractions is a way of expressing numbers that are not whole using divisions of tenths, hundredths, thousandths, etc.

Denominator: The lower number in a fraction; the lowest common denominator is the figure to which all denominators can be converted. For example the lowest common denominator of 1/2 and 1/3 is 6 (3/6 and 2/6).

Digit: Another word for a numeral, a single number. Digit is also used as a biological term for fingers or toes—which were perhaps the first things humans ever counted.

Gradient: A measure of the slope of a line, which is based on the proportionality of x and y; if x = y the gradient is 1, if 3x = y, then the gradient is 3; the value of x increases by a factor of 3 for every 1 of y.

Infinity: A limitless, endless number that can never be reached. The fraction 1/x tends to (get closer to) infinity as x tends to (get closer to) zero.

Integer: A whole number, positive or negative, therefore not including the fractions. Zero is generally included with the integers, despite it being neither whole nor fractionable.

Inverse: To change the sign of a number, from positive to negative or vice versa—the inverse of 3 is –3.

Logarithm: A counting system that uses the exponent of a number rather than the number itself; log 3 of 9 is 2 (3^2=9), while $3\log_3$ is 27 (3^3).

Matrix: An array of numbers set out in rows and columns, which can be added, multiplied, etc, en masse. Matrices are used to express multiple values that are grouped or linked in some way, such as the dimensions of a shape. The matrix can be used to calculate the new dimensions of the shape after it has been transformed by some means.

Mean: The most commonly understood average, calculated by dividing the total of the sample by the number of samples taken, therefore the mean of 2, 4 and 6 is 12/3 = 4, while the mean of 2, 3, and 5 is 10/3=3.333...

Median: An average that is taken by finding the middle figure in a sample, which is larger than the lower half, but smaller than the upper half. The means of 1, 3, and 8 and 2, 4, and 6 are both 4, while the medians are 3 and 4 respectively.

Mode: An average figure that is the most common number occurring in a sample. The mean of 1,1,4,7 is 3 but the mode is 1.

Ordinal number: Numbers that indicate order, first, second, third, etc.

Radix point: The point used in a number to denote when the value is being counted in fractions of the radix, or base. Generally this is better known as a decimal point because the radix we all use is 10. (In Europe the point is frequently written as a comma.) So in base 10, 10.1 is the decimal fraction of 10 1/10 or ten and a tenth. In base two (binary) 10.1 is 2 1/2 or two and a half.

Reciprocal: The number that is derived by dividing another into 1. The reciprocal of 4 is 0.25 (a quarter).

Reoccurring: When a decimal fraction has an infinite string of decimals of a single number: a third (1/3) is 3.3 reoccurring.

Scalar: A quantity that has only one value, for example, speed.

Vector: A quantity that has two or more values, for example velocity, which is speed and direction.

$$6 = 1 + 2 + 3$$

$$28 = 1 + 2 + 4 + 7 + 14$$

$$496 = 1 + 2 + 4 + 8 + 16 + 31 + 62 + 124 + 248$$

$$8128 = 1 + 2 + 4 + 8 + 16 + 32 + 64 + 127 + 254 + 508 + 1016 + 2032 + 4064$$

PERFECT NUMBERS

A perfect number is any number that is the sum of its divisors. For example, 6 can be divided by 1, 2, and 3, and these numbers add up to 6. Perfect. Only 47 perfect numbers are known so far. The largest has 25956377 digits. They are all even numbers. No one knows if any odd numbers can be perfect but the search continues.

Operations

A MATHEMATICAL OPERATION IS A PRE-DEFINED PROCEDURE THAT CONVERTS ONE OR MORE INPUTTED NUMBERS into one or more outputs. It is not just adding up and multiplication tables.

FACTORIALS

Numbers grow amazing quickly as factorials.

0	1
×1	1
×2	2
×3	6
×4	24
×5	120
×6	720
×7	5040
×8	40320
×9	362 880
×10	3 628 800
×11	39 916 800
×12	479 001 600
×13	6 227 020 800
×14	87 178 291 200
×15	1 307 674 368 000
×18	6 402 373 705 728 000
×20	2 432 902 008 176 640 000
×24	6 2044 840 200 000 000 000 000 000 000 000

Addition: Starting at the beginning, addition is combining numbers together to form a single larger one. The result of an addition is called the sum.

Division: The number of times a smaller number fits into a larger one. Any figure left over is the remainder. The opposite of multiplication.

Exponentiation: An operation that involves a figure being raised to the power of another; 10^n involves multiplying 10 by itself n times.

Multiple: A multiple of a number is another that can be arrived at by multiplying the former. For example, 6 is a multiple of 2 (2 x 3).

Multiplication: The operation that involves adding a number together a multiple of times. For example, 10 x 3 = 10 + 10 + 10. The result of a multiplication is called a product.

Subtraction: To find the difference between two numbers, taking one from the other.

Cube: Used to describe the exponent 3, two cubed is eight ($2^3 = 8$).

Divisor: A number that divides exactly into another, also called a factor.

Exponent: The power to which another number is raised, usually written in superscript.

Factor: Another word for divisor.

Factorial: The product of all positive whole numbers less than or equal to a specific number. Four factorial, written 4!, is 1 x 2 x 3 x 4 = 24.

Product: The result of a multiplication.

Quotient: The result of a division.

Square: To raise a number to the power 2.

Square root: The number that has to be squared to produce another. The square root ($\sqrt{}$) of 4 is 2 ($2^2 = 4$).

Transformation: When a set of numbers is inputted into the same function, and transformed in the same way. For example, a simple function of halving all lengths would transform a 2 by 2 square into a 1 x 1 square.

Geometry

THE MATH OF SHAPES AND FORMS DOES NOT JUST USE NUMBERS TO DESCRIBE THEM, but also uses many words to sum up the characteristics of objects. We are all familiar with squares, triangles, and angles but there is a lot more going on.

Acute: When an angle is less than a right angle (90°).

Arc: A section of a circumference.

Area: The region covered by a surface measured in squared units.

Chord: A straight line drawn from one point on a circumference to another. The diameter is the chord that passes through the center of the circle.

Circumference: The perimeter of—line that surrounds—a circle.

Congruent: In geometric terms, congruent shapes are the same shape and size—and can cover one another exactly.

Cuboid: A six-sided solid made from quadrilaterals; a cuboid made of six squares is a cube.

Degree: A unit of angle; a complete turn is 360°.

Diameter: The distance from one side of a circle to the other through the center. The diameter is twice the radius.

Equilateral: Of equal side length; often referring to a regular triangle with equal sides (and equal angles).

Geodesic: The shortest distance between two points on a sphere or other curved surface.

Isosceles: A triangle that has two equal sides and one unequal one.

Obtuse: An angle that is greater than a right angle (90°).

Parallel: A line that will never intersect with another if extended in any direction.

Perpendicular: When a line intersects another at 90°.

Plane: A flat surface where the shortest distance between two points is a straight line.

Point: A dimensionless location in space.

Polygon: A flat shape made from straight lines. There are an infinite number of regular polygons, with sides of equal length.

Polyhedron: A three-dimensional solid made from straight lines and flat faces. There are just five regular polyhedra.

Polytope: A geometric shape with straight lines and flat faces in four or more dimensions.

Regular Pentagon

Sides	5
Each angle	108°
Sum of angles	540°
Number of triangles	3
Diagonals	5

Equilateral Triangle

Sides	3
Each angle	60°
Sum of angles	180°
Number of triangles	1
Diagonals	0

Regular Hexagon

Sides	6
Each angle	120°
Sum of angles	720°
Number of triangles	4
Diagonals	9

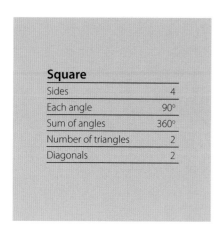

Square

Sides	4
Each angle	90°
Sum of angles	360°
Number of triangles	2
Diagonals	2

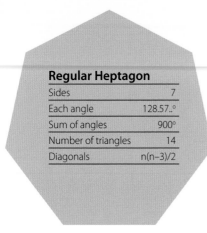

Regular Heptagon	
Sides	7
Each angle	128.57..°
Sum of angles	900°
Number of triangles	14
Diagonals	n(n–3)/2

Radian: A unit for measuring angles; one radian is the angle of a arc with a length of one radius. A full circle is 2π radians.

Radius: The distance from the center of a circle to any point on the circumference.

Regular: A shape that has sides of equal length, angles of equal size, and equal areas.

Right angle: An angle of 90 degrees.

Scalene: A triangle that has no equal sides or angles.

Regular Polygon	
Sides	n
Each angle	(n–2) x 180°/n
Sum of angles	(n–2) x 180°
Number of triangles	(n–2)
Diagonals	n(n–3)/2

Similar: In geometric terms, similar shapes have the same angles and proportions but are not the same size.

Symmetry: A property of a shape (or other group of numbers) that can be transformed—for example, rotated or reflected—and still appear the same.

Tangent: A line that passes through just one point of a curve such as a circle.

Regular Octagon	
Sides	8
Each angle	135°
Sum of angles	1080°
Number of triangles	6
Diagonals	20

Vertex: The name for the point where edges or other lines meet. A polygon, a polyhedra, and graphs are collections of vertices.

Volume: The space filled by a three-dimensional object.

Expressions

FINALLY, A QUICK LOOK AT MATHEMATICAL EXPRESSIONS, the means by which mathematicians present their findings with symbols.

Algebra: Investigating the relationships between numbers by replacing real, yet variable, numbers with generalized terms, most frequently x and y.

Coefficient: A multiplicatory value in an expression that is non-variable. In $c = 2\pi r$, 2 is the coefficient.

Constant: An interesting unchanging number that appears in formula; π is a mathematical constant.

Equality: When two things are the same.

Equation: Showing that one mathematical expression equals another: $2x = 3y$.

Formula: An equation that is used to calculate a specific value. The formula for the area of a circle is Area = $\pi r2$.

Function: A defined process that converts input values into outputs using one or more operations.

Identity: An algebraic expression that equals one.

Inequality: When two things are not the same.

Types of Proof

To DISPROVE A THEOREM A MATHEMATICIAN NEED ONLY PROVIDE A COUNTEREXAMPLE, an exception that disproves the rule. While this shows a theorem is not universal, it may still be true within certain limits, which is the next problem to solve.

For a mathematical theorem to be accepted as true it must have a proof. The proof is a demonstration that the theorem is correct in every case or within certain limits. Sometimes a proof is known as a lemma, which is a proof that is a stepping stone to a more complete final proof. There are several types of proof, each of which are based on axioms which are assumed to be true and do not themselves require proof.

Direct proof: Applying the axioms and any other definitions to deduct a universal truth.

By mathematical induction: By solving the problem in one or few cases and then showing that it must be true in all other cases (or within limits).

By transposition: Showing problem A is true by proving that problem not A has an opposite result.

By contradiction: Proving that a statement is logically contradictory, thus false, and so proving that the valid statement is the opposite (or only possible alternative) to the original statement. Also known as *reductio ad absurdum* (reduce to the absurd).

By construction: Showing that a property is true by using it to construct an exemplar that exhibits that property.

By exhaustion: To perform a large number of calculations without finding a contradictory result.

Probabilistic proof: Proving with an example of the property by using probability to show that the example is certain to exist.

Combinatorial proof: Proving one thing by comparing it to a known mathematical object and showing the two are equivalent.

Nonconstructive proof: Establishing that a mathematical property must exist even though it is not possible to explain how to isolate it.

Statistical proofs in pure mathematics: Using statistics to show that a solution to a problem is true or not.

Computer-assisted proofs: Using the processing power to perform an exhaustive proof, and self-checking the calculations that are way beyond the scope of a human mathematician.

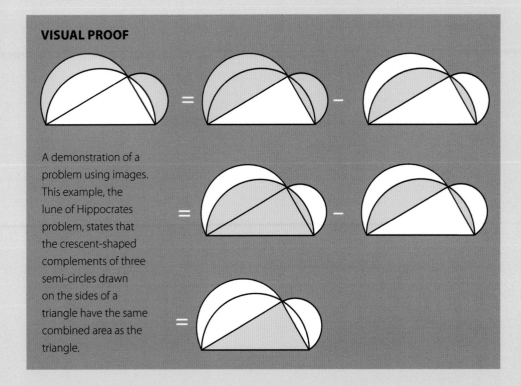

VISUAL PROOF

A demonstration of a problem using images. This example, the lune of Hippocrates problem, states that the crescent-shaped complements of three semi-circles drawn on the sides of a triangle have the same combined area as the triangle.

A Number of Numbers

NUMBERS ARE INFINITE, BUT THERE ARE SEVERAL SETS OF INFINITE NUMBERS that nest among each other. There is an infinite number of these sets, but we will start at the beginning.

6. Imaginary numbers: Imaginary numbers are born out of the square root of –1. Real square numbers are always positive (two alike signs make a +). So while the square root 1 is 1 (the unit of real numbers), the square root of –1 is *i* instead, and this is the unit used in imaginary numbers. Like the real numbers, imaginary numbers can be plotted along a number line, or axis, it is just a different one. The imaginary number line meets the real line at just one point—zero. Anything multiplied by zero equals zero, even *i*: $0 = 0i$. *i* may not be real but it obeys this rule and all the others nonetheless, and that means some imaginary numbers are algebraic, while some are transcendental.

7. Complex numbers: The word complex refers to numbers that have two parts, one real, one other imaginary. This set of numbers contains all sets of all other numbers. But does the complex set fit within a larger one?

1. Natural Numbers: The numbers we count with using our fingers, then toes—and then with the power of our minds. If we had the time we could count them to infinity. However, if we did we would be missing out on an infinity of other numbers.

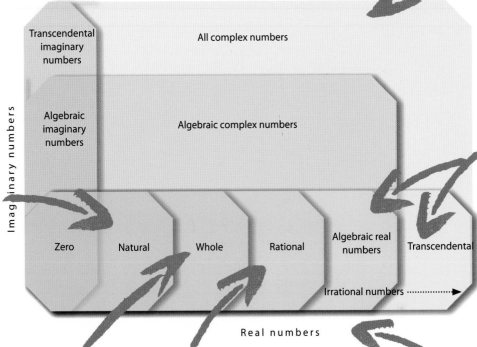

4. Irrational numbers: As you might expect, the irrational numbers are not rational, and cannot be expressed as a fraction. Instead they form an infinite stream of decimal places with no repeating pattern. Some irrational numbers can be constructed into algebraic expressions, and so are called algebraic or constructible numbers. The irrationals that defy being expressed as algebraic solutions are known as transcendental numbers.

2. Whole Numbers: Also known as integers, this set contains the natural numbers but also includes the negative numbers, and zero gets in there too. Negative numbers are identical in every way to natural number only they just get less as they get larger.

3. Rational Numbers: These are numbers that can be expressed as fractions, which are always constructed from one integer divided by another: $1/2 = 0.5$; $7/4 = 1.75$; $97/29 = 3.34482759$. If one of the integers is negative, then the rational number is also negative. (In multiplication and division, like signs produce positive results; unlike signs result in negative numbers.) Any whole number can also be expressed as a fraction $2/2 = 1$; $9/3 = 3$; $64/4 = 16$.

5. Real numbers: The irrationals, and rationals within, form the real numbers. Why real? Well because there is a whole other set that runs in the same way, it is just its members do not exist. But that needn't stop us, instead of counting the 1s we count in *i* for *imaginary*.

To infinity and beyond...

It does not end with the complex numbers. That is really only the beginning. A complex number has a real component x and another in the form of i. Imagine all real numbers form a point somewhere along an x axis of a graph, well the imaginary numbers run perpendicular to it on a z axis. Why not add two more sets of numbers, j and k, two more number lines in two more dimensions. The complex numbers with four parts are called quaternions, and the rules on how the four number sets relate is shown in this table. It goes on; octonions use eight number sets... and, you guessed it, the possible number is infinite.

multiply	1	i	j	k
1	1	i	j	k
i	i	-1	k	j
j	j	k	-1	i
k	k	j	i	-1

COUNTING SYSTEMS

Decimal	Roman	Hexadecimal	Binary
1	I	1	1
2	II	2	10
3	III	3	11
4	IV	4	100
5	V	5	101
6	VI	6	110
7	VII	7	111
8	VIII	8	1000
9	IX	9	1001
10	X	A	1010
50	L	32	110010
100	C	64	1100100
500	D	1F4	111110100
1000	M	3E8	1111101000

GREEK SYMBOLS

		Name	Sound
Α	α	Alpha	a
Β	β	Beta	b
Γ	γ	Gamma	g
Δ	δ	Delta	d
Ε	ε	Epsilon	e
Ζ	ζ	Zeta	z
Η	η	Eta	h
Θ	θ	Theta	th
Ι	ι	Iota	i
Κ	κ	Kappa	k
Λ	λ	Lambda	l
Μ	μ	Mu	m
Ν	ν	Nu	n
Ξ	ξ	Xi	x
Ο	ο	Omnicrom	o
Π	π	Pi	p
Ρ	ρ	Rho	r
Σ	σ	Sigma	s
Τ	τ	Tau	t
Υ	ψ	Upsilon	u
Φ	φ	Phi	ph
Χ	χ	Chi	ch
Ψ	ψ	Psi	-
Ω	ω	Omega	-

THE STORY OF MATHEMATICS WILL NEVER BE OVER, WITH AN INFINITY OF PATTERNS, RELATIONSHIPS, AND LAWS TO DISCOVER. Here are a few open questions that still need an answer, and there are dozens more besides. And once resolved these problems will inexorably lead to more mysteries that need to be solved.

Is there such a thing as a perfect box?

Only in math could this question hold any interest. The story begins with Euler bricks, a class of cuboids where the lengths of the sides and the diagonal lines across all faces have integer lengths—they are all whole numbers, no infinite decimals or even fractions here. The smallest Euler brick has sides of 240, 117, and 44 units and diagonals of 267, 244, and 125 units. So a perfect box is a Euler brick that also has integer values for its space diagonals—the lines that run through the middle of the shape between opposite corners. In 2009 perfect parallelepipeds (wonky boxes) were shown to exist but as yet the perfect box has proved elusive. It is not the case that nobody is looking: as of 2012, computational analysis of the problem has ruled out all boxes with a longest side less than one trillion units.

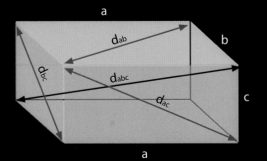

Is it possible to idealize foam?

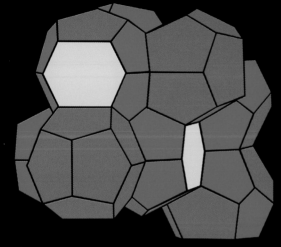

One should not take this question too literally, but instead view it in terms of math. Ideal foam is the answer to a question posed by Lord Kelvin, he of the laws of thermodynamics, way back in 1887. Having completed his work on the nature of energy that had assuredly immortalized his name for centuries to come, the Scottish peer idled away his dotage by thinking about the shapes of bubbles in foam. If all the bubbles were the same volume what is the most efficient shape for packing them together? Kelvin suggested a 14-sided truncated octahedron (with its pointy bits cut off to leave square facets). These shapes packed together formed what became known as the Kelvin structure. Then in 1993 two Dublin-based researchers bettered the lord—twice. Denis Weaire and his student Robert Phelan showed that a tetrakaidekahedron, a shape made up of two hexagons and 12 pentagons, packed together with a lower area of touching surfaces. Secondly an irregular dodecahedron was shown to share this lower surface area— albeit 0.3 per cent less than the Kelvin structure. As is often the case, nature got there first. The clathrates, a class of organic molecules that trap others in their structure, have members that approximate to the Weaire-Phelan structure.

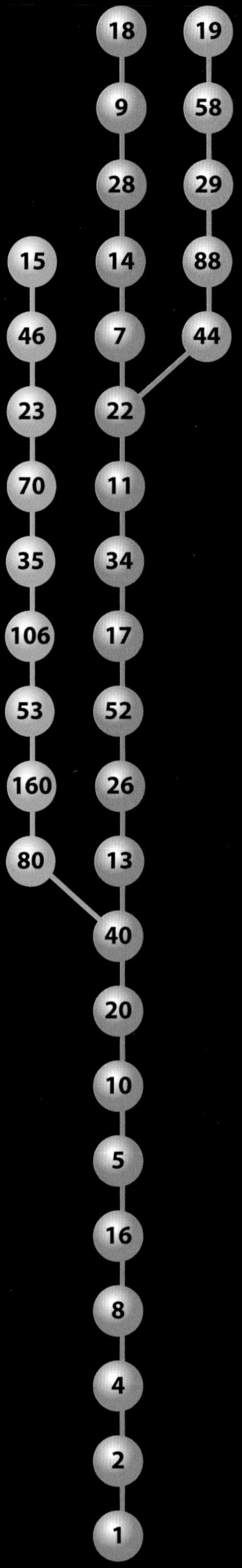

The Collatz conjecture

Try this one at home—if you dare. Named for Lothar Collatz in the 1930s, this unsolved problem relies on the HOTPO process. This stands for "half or triple plus one" and is a simple algorithm that a child could learn, although no mathematician has yet mastered it. The rules are, pick a number (x), if it is even half it ($x/2$), if it is odd triple it and add one ($3x + 1$). What next? Keep repeating HOTPO and Collatz suggests you will always arrive at the number one. However, is this always the case? Is there a number where oneness is not achieved? The late Paul Erdös, a contemporary of Collatz, offered $500 to anyone who found out. That prize may not be worth so much 80 or so years later but a proof either way could reveal a new pattern connecting natural numbers—and who knows where that would lead.

Collatz sequence diagram:
18 · 19
9 · 58
28 · 29
15 · 14 · 88
46 · 7 · 44
23 · 22
70 · 11
35 · 34
106 · 17
53 · 52
160 · 26
80 · 13
40
20
10
5
16
8
4
2
1

Is there a pattern in the primes?

Mathematicians have ceaselessly sought patterns within prime numbers in the hope that they may reveal a hidden formula of some kind. The mathematical power won by anyone who has mastered the primes is so far the stuff of science fiction, but then so were space travel, submarines, and robots once. A pattern proposed by Alphonse de Polignac in 1849 was this: there are infinitely many cases of two consecutive prime numbers with difference n, where n is a positive even number. Putting it another way, pick any even number and there will be an infinity of primes that you can add to it that will result in another prime but no other prime number will fall in between. Primes with a gap of 2 are called twins, those with a gap of 4 are cousins, while a prime gap of 6 makes them sexy primes (from the Latin for *six*, rather than anything untoward).

5 6 **7**
Twins

7 8 9 10 **11**
Cousins

11 12 13 14 15 16 **17**
Sexy

Goldbach's conjecture

In 2000, London publisher Tony Faber offered a cool $1 million to anyone who solved this most venerable of math puzzles. (At the same time the Clay Foundation was setting out its goals for the coming century.) But the publisher put a time limit of two years on the prize, counting from the release date of the English edition of a novel, *Uncle Petros and Goldbach's Conjecture*. With or without publicity stunts, this 1742 conjecture has continued to stump the brightest and the best, even the great Leonhard Euler who was seldom foiled. The proposal made by Christian Goldbach was: every even number greater than 2 is a composite of just two prime numbers. Every even number (except 2) is the sum of two prime numbers. No one has found anything to disprove this assertion by exhaustive additions, and a flimsier sibling, the "weak Goldbach conjecture" has joined the family: all odd numbers greater than 7 are the sum of three odd primes. Both the strong and weak conjectures remain highly likely but as yet there is no proof either way.

100 = **3+97**

or **100 =** **11+89**

or **100 =** **17+83**

or **100 =** **29+71**

or **100 =** **41+59**

or **100 =** **47+53**

Finding the way

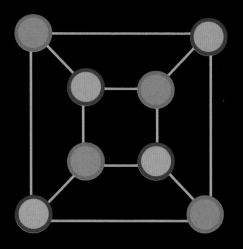

Barnette's conjecture takes us to the faintly odd world of graph theory. Named for David Barnette, a professor from the University of California, this problem is looking for a universal law about finding a route around a certain type of graph. In this context the graph is not a chart displayed between axes, more a network with lines, or edges, running between nodes, or vertices. The problem says that every bipartite polyhedral graph with three edges per vertex has a Hamiltonian cycle. Let's try that again a little slower. In graph theory, a polyhedral graph represents the edges and vertices of a 3D shape but presented on a flat plane. Bipartite indicates that the vertices are allotted one of two colors, with each edge beginning and ending with a different color. The conjecture suggests that every graph such as these has a route through it where each vertex is passed once only—a so-called Hamiltonian cycle. So far it has been found that every graph up to 86 vertices complies.

The Happy Ending problem

This is another example of a math problem that is literally child's play. The problem was proposed by Esther Klein at an informal gathering of young mathematicians hanging out in pre-war Budapest: any five points have a subset of four that form a convex quadrilateral. Two young men present, George Szekeres and Paul Erdös, wondered why this was. They found that nine points always contained a pentagon and 17 had a hexagon, but no one has proved why this is always true. Young George perhaps had more reason than most to be interested in Esther's puzzle—the couple married three years later, hence the "happy ending."

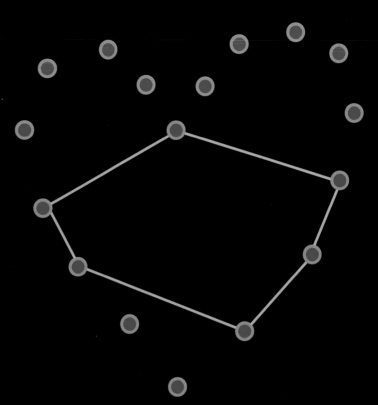

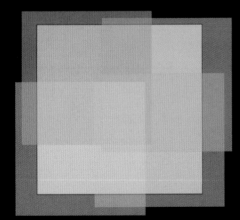

Covering shapes

Hugo Hadwinger has several conjectures named for him. This one relates to combinatorial geometry, specifically how many smaller but similar shapes it takes to cover a single larger one. Similar shapes have the same geometry, but are just scaled up or down. It takes three similar triangles to cover their larger version; while a square needs four. Hadwinger's conjecture says the number needed is always a maximum of $2n$ where n is the number of dimensions. Hence a 2D square needs four copies to cover it, a 3D cube would need six smaller cubes to occupy its entire space. The 2D case is solved but dimensions 3 and above are still unproven, leaving open the possibility of strange shapes that defy covering or defy scaling.

The Great Mathematicians

FOR MANY OF US, COMPLEX MATH IS BEYOND OUR COMPREHENSION and the people that can explore its infinite landscape of numbers and patterns are rightly held in wonder. However, they are in many ways ordinary people, but they all have an extraordinary story to tell.

Archimedes

Born	c.290–280 BC
Birthplace	Syracuse, Sicily
Died	212/211
Importance	First accurate calculations of pi

It is unknown whether this Greek scientist and engineer ever went to what we now understand as Greece. Instead he lived in the Hellenic colony of Syracuse in Sicily, and may have studied in Alexandria under Eratosthenes. Few people have achieved as much as Archimedes. Aside from the math, he invented the Archimedes screw for lifting water; the claw of Archimedes, a battle crane used for sinking ships; a mysterious heat ray weapon, which is thought to be intense beams of sunlight focused by mirrors; and of course the Archimedes principle, which relates buoyancy to density.

Al-Khwarizmi

Born	c.780
Birthplace	Khorasan, central Asia
Died	c.850
Importance	Inventor of algebra

Known as Algoritmi by Latin-speakers, this Persian scientist was probably born in Khorasan, south of the Aral Sea. In the 9th century this was a wealthy outpost of the burgeoning Islamic empire but it is now part of Uzbekistan, where al-Khwarizmi is still hailed as a national hero. The center of learning in his day was Baghdad, and al-Khwarizmi became a scholar at the city's House of Wisdom. As well as extending algebra and algorithms, al-Khwarizmi is remembered for his designs of sundials and astrolabes (star mapping devices) and for drawing up one of the most accurate world maps of his time.

Babbage, Charles

Born	December 26, 1791
Birthplace	London, England
Died	October 18, 1871
Importance	Inventor of mechanical computer

As a student Babbage is reported to have been disappointed with the math teaching at Cambridge University, and, inspired by the work of Leibniz and Lagrange, set up the Analytical Club with John Herschel (of the astronomical family) and others. As was quite normal for men of science of his day, Babbage was also a leading figure in the Ghost Club, which investigated the supernatural. He decided to design a mechanical computer because of all the errors made by human mathematicians. However the sheer number of precision gears required made the device prohibitively expensive.

Bayes, Thomas

Born	1702
Birthplace	London,England
Died	April 17, 1761
Importance	Advances in statistics and probability

The Reverend Thomas Bayes spent most of his working life tending to a congregation at his Presbyterian chapel in Tunbridge Wells, a market town to the southeast of London. In his 30s, Bayes began to dabble in mathematics. He was elected to the Royal Society in 1742 for his work on calculus, which was still very much a new area of math, with the work of Newton and Leibniz only decades old. However, the probability theory for which he is now remembered was born out of an interest in the last few years of his life. The manuscript on Bayes' theorem was not published until after his death.

Boole, George

Born	November 2, 1815
Birthplace	Lincoln, England
Died	December 8, 1864
Importance	Invented Boolean algebra and logic

This English logician's genius was apparent early in life. He grew up in a meager home, given extracurricular schooling by his father and friends of the family, but largely teaching himself, learning several languages from books and even mastering calculus (eventually). By the tender age of 16 the young George was a teacher himself, becoming the best-paid member of the family. Boole's success as a teacher did not go unnoticed and in 1849 he was made the first math professor at a new university in Cork, Ireland, which is where he completed his work on symbolic logic.

Bernoulli, Daniel

Born	February 8, 1700 (Jan. 29, Old Style)
Birthplace	Groningen, Netherlands
Died	March 17, 1782
Importance	Developed field of fluid dynamics

The Bernoullis of Basel, Switzerland were a mathematical family like no other. Daniel's uncle Jacob discovered the value of e, while his father Johann made contributions to calculus and solved the brachistochrone, the curve taken

when an object swings from one place to another (it is not quite circular). Daniel himself is remembered for the math of fluid dynamics. His is the name attributed to the principle of lift that gets aircraft off the ground. Nevertheless his cousins, brother, and several nephews and great-nephews were also famous mathematicians.

Cantor, Georg

Born	March 3, 1845
Birthplace	St. Petersburg, Russia
Died	January 6, 1918
Importance	Founder figure of set theory

Despite being born in St. Petersburg, the jewel of the Russian Empire, Cantor was an ethnic German, whose parents were migrant traders. George moved to Germany at the age of 11, after his parents despaired of the Russian climate. George excelled at school and became the Extraordinary Professor of Math at Berlin University when still only 34.

Soon his set theory put him at the center of world math, but in his 50s Cantor began a struggle with depression that often resulted in hospitalization. His retirement was spent in poverty, struggling to feed himself during the First World War.

Descartes, René

Born	March 31, 1596
Birthplace	La Haye, Touraine, France
Died	February 11, 1650
Importance	Inventor of planar coordinates

Few phrases are more quoted than "Cogito ergo sum," René Descartes' pithy proof of his own existence. Descartes' fuller explanation is that where there is thought, there is also doubt. If one doubts one's thoughts, one's existence, that is proof enough that you exist. A staunch Roman Catholic, Descartes chose to live in Dutch territories, where Protestantism was to the fore. He shelved his first major work, *Treatise on the World*, after his contemporary Galileo was put on trial by the Vatican for heresy. However, much of this work later found its way into his masterwork *Discours de la méthode*.

Euclid

Born	flourished 300 BC
Birthplace	Alexandria, Egypt
Died	-
Importance	Author of *Elements of Geometry*

The life of Euclid is a mystery. Many translations of his work, including the first in English, attribute it to one Euclid of Megara, who was a pupil of Socrates who probably lived a century before our Father of Geometry. The mathematical Euclid was based in Alexandria in the 4th and 3rd centuries BC where he undoubtedly consulted the contents of the recently set up Great Library. Archimedes reports that he was the tutor of Ptolemy, Egyptian pharaoh, and admonished the king for trying to cut corners, saying "There is no royal road to geometry."

Eratosthenes

Born	c.276
Birthplace	Cyrene, Libya
Died	c.194 BC
Importance	Calculated the size of Earth

As the chief librarian at the Great Library of Alexandria, Eratosthenes had the largest information resource the world had ever seen at his fingertips, and he put it to use during his famous measurement of the globe. This feat among others earned Eratosthenes the title of founder of geography, a word he himself coined. The scientist was also a noted early egalitarian, criticizing Aristotle's remarks that Greek blood should be kept pure by avoiding marriage with the "barbarian" peoples. Being of North African origin himself, Eratosthenes would probably not have passed Aristotle's heritage test.

Euler, Leonhard

Born	April 15, 1707
Birthplace	Basel, Switzerland
Died	September 18, 1783
Importance	Founder of graph theory

As well as having one of the most mispronounced names in math (it is something similar to "oiler"), Euler was a founding figure in graph theory, natural logarithms, and infinitesimal calculus, as well as dabbling in logic, optics, and structural engineering. Swiss by birth, Euler was friendly with the second generation of the Bernoulli family, sometimes tutored by the first. He worked mainly in St. Petersburg and Berlin, and his achievements are all the more remarkable given he spent more than half his life with very poor eyesight,

Fermat, Pierre de

Born	August 17, 1601
Birthplace	Beaumont-de-Lomagne, France
Died	January 12, 1665
Importance	Famous for his "Last Theorem"

Perhaps the greatest amateur mathematician the world has ever known, Pierre de Fermat was a lawyer in the sleepier towns of southern France—and by all accounts he was not universally admired for his courtroom skills. A private and cautious man, Fermat chose relative obscurity in a time when public life in France was a risky business, with religious violence frequently flaring up. Fermat published little and much of his work has been pieced together after his death. He left few proofs (it was not *de rigueur* in those days), and his most famous theorems were proven by others.

Fourier, Joseph

Born	March 21, 1768
Birthplace	Auxerre, France
Died	May 16, 1830
Importance	Developer of the math behind complex waves

Raised by nuns, the orphaned Fourier was blocked from the French army's scientific corps, which was reserved for the sons of noblemen, and took a military lectureship in mathematics, concerned largely with the math of ballistics. After the Revolution Napoleon, himself a former artillery officer, made Fourier the governor of Egypt. Fourier later showed the Rosetta Stone to Jean-Francois Champollion, who deciphered its hieroglyphics. Fourier was the first to describe how the energy in sunlight is trapped by the atmosphere—the so-called Greenhouse Effect.

Fibonacci

Born	c.1170
Birthplace	?Pisa, Italy
Died	after 1240
Importance	Defined the Fibonacci series

Also known as Leonardo of Pisa or Leonardo Bonacci, this mathematician's more familiar moniker is a patronymic pet name, meaning "son of Bonacci". The name we use for him today (and his arithmetical progression) only arrived in common usage long after his death. Leonardo's mathematics sprung from a childhood spent in what is now Algeria. (His merchant father ran a Pisan trading post at Bugia.) In adult life, Leonardo was salaried by the Republic of Pisa in recognition of his math work, which was seen largely as a means to improve business practices.

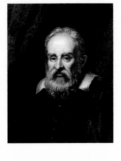

Galileo

Born	February 15, 1564
Birthplace	Pisa, Italy
Died	January 8, 1642
Importance	Defined laws of fall and pendulums

This scientist is known best as an astronomer and physicist but he was among the first to apply math to their investigation. The son of a musician-cum–mathematician, Galileo chose a career in science, but was always on the look-out for a business opportunity— his family always had money troubles. The telescope was one such get-rich-quick scheme, earning him various pensions. However, his heliocentric description of the Universe that he saw through his telescope put him in conflict with the church, and to avoid jail and secure his income, Galileo was forced to recant.

Gauss, Carl Friedrich

Born	April 30, 1777
Birthplace	Brunswick, Germany
Died	February 23, 1855
Importance	Leading figure in several fields

The so-called "Prince of Mathematics" had a decidedly unroyal start in life. His parents were illiterate and did not record his birth. His mother recalled he was born on the Wednesday eight days before Ascension, itself 40 days after Easter. Gauss devised a method for calculating Easter of any year, past or future, to track down his own birthday. A math prodigy at school, Gauss's education was sponsored by the Duke of Braunschweig, who sent him to the college in Göttingen, where he was based for the rest of his life. Gauss was the towering figure of his generation, making contributions to geometry, prime numbers, and statistics.

Hamilton, William

Born	August 4, 1805
Birthplace	Dublin, Ireland
Died	September 2, 1865
Importance	Discovered quaternions

This Irish mathematician and astronomer made his mark quite literally, by scratching his formula for quaternions into the stone of a Dublin bridge. As a child he was already showing prodigious talent for math. When Zerah Colburn, an American savant gave a show in Dublin (performing complex calculations at great speed), a 12-year-old Hamilton took him on, with some success. He also learned several languages as a child. Hamilton spent his academic career at Trinity College, Dublin, where he worked on everything from optics to a reformulation of the laws of motion.

Gödel, Kurt

Born	April 28, 1906
Birthplace	Brünn, Austria-Hungary (now Brno, Czech Rep.)
Died	January 14, 1978
Importance	Developed incompleteness theorem

An ethnic German, born in the Czech city of Brno (then part of the teetering Austro-Hungarian empire), the young Kurt went to Vienna to study. It was in this city where, at the age of 25, he published the incompleteness theory that came to define him. A few years later Nazis killed his Jewish mentor, and Gödel had a nervous  collapse. When the war started a few years later Gödel fled to Princeton U.S.A. under the encouragement of his friend Albert Einstein. Gödel suffered metal illness for the rest of his life. He only ate food prepared by his wife, and when she was hospitalized, he refused to eat and starved to death.

Hilbert, David

Born	January 23, 1862
Birthplace	Königsberg, Prussia (now Kaliningrad, Russia)
Died	February 14, 1943
Importance	Proposed 23 problems for the 20th century

This German mathematician is best remembered perhaps for the great challenges he made to his colleagues at the dawn of the 20th century. Hilbert was, however, not just a great promoter and teacher of math, but one of its greatest discoverers, too. He was an East Prussian by birth, and spent his adult career at Göttingen, the alma mater of Gauss. After his retirement, the Nazis purged Jews from the Göttingen faculty, leading Hilbert to complain to the new minister of education that the study of mathematics had ended at the university altogether.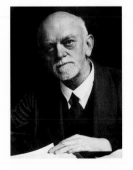

Hipparchus

Born	?
Birthplace	Nicaea, Bithynia (now Iznik, Turkey)
Died	after 127 BC
Importance	Developed trigonometry

Perhaps better known as an astronomer, Hipparchus developed what would become the field of trigonometry in order to explain the motion he observed in heavenly bodies. Hipparchus spent much of his life on the Aegean island of Rhodes (off the Turkish coast but still part of Greece). His hunch was that the planets moved around the Sun, and he was the first to calculate their motion. However, the results indicated that the planets did not move in perfect circles, enough for Hipparchus to abandon the idea as manifestly incorrect: the universe was perfect and so therefore must be its motion.

Leibniz, Gottfried

Born	July 1, 1646 (June 21, Old Style)
Birthplace	Leipzig, Germany
Died	November 14, 1716
Importance	Co-developer of calculus

By all accounts Leibniz was the complete opposite of his rival Isaac Newton: he was humorous and charming with admirers all over Europe. He came to math late in life. He was a diplomat, first for the Elector of Mainz then and, as the conflict over calculus with Newton was festering, Leibniz took employment with George of Hanover, soon elevated to king of Great Britain. That made Leibniz a powerful advisor of Newton's monarch, yet his new influence had little effect on the argument. His high place resulted in a meteoric fall, and Leibniz died in obscurity.

Laplace, Pierre Simon

Born	March 23, 1749
Birthplace	Beaumount-en-Auge, Normandy, France
Died	March 5, 1827
Importance	Leading figure in several fields

From aristocrat to imperial scientist, few have lived through such change as this French scientist who had a hand in so many scientific developments. He dabbled in thermodynamics with his soon-to-be guillotined friend Antoine Lavoisier. He was among the first to propose that the Solar System had developed from a nebula of hot, swirling gas; and when asked by his emperor Napoleon Bonaparte why God was not mentioned in his work, Laplace responded, "I have no need of that hypothesis." In terms of math, he did fundamental work on probability, statistics, and mechanics.

Mandelbrot, Benoit

Born	November 20, 1924
Birthplace	Warsaw, Poland
Died	October 14, 2010
Importance	Leading figure in fractal geometry

Mandelbrot was among the first of a new breed of mathematicians who harnessed the power of computing to push back the boundaries of the subject. Mandelbrot's early years were punctuated by relocations forced by the Nazi threat. He left Warsaw for Paris at the age of 11 and then headed to the relative safety of Vichy France during the Second World War. His early work was in a wide variety of applied fields of math, and Mandelbrot became fascinated by the self-similar structures that appeared in them all, an interest that led him to fractal geometry.

Napier, John

Born	1550
Birthplace	Merchiston Castle, near Edinburgh, Scotland
Died	April 4, 1617
Importance	Invented logarithms

Known as the eighth Laird of Merchiston, John Napier was an eccentric Scottish aristocrat who lived hidden away in a castle—now part of Napier University in Edinburgh—seldom seen outside without his trademark black cloak and black rooster. Napier cultivated a reputation for being a wizard. One story goes that he exposed a thief among his staff by asking them one by one to stroke his rooster (blackened with soot)—the magic chicken, he said, would mark the guilty's hand. The innocent with nothing to hide stroked the bird, while the culprit did not, and was revealed by his clean hands.

Oughtred, William

Born	March 5, 1574
Birthplace	Eton, Buckinghamshire, England
Died	June 30, 1660
Importance	Invented slide rule

Among many innovations in mathematics, we have William Oughtred to thank for the x multiplication symbol and the slide rule. Perhaps fewer of us would thank him for introducing the abbreviation sin and cos for use in trigonometry. In keeping with the long tradition of men of

knowledge, Oughtred was a clergyman by profession and had a strong interest in astrology and the occult. In later life he became a teacher, counting among his students Christopher Wren, the architect of London's St Paul's Cathedral, the Royal Observatory at Greenwich, and Oxford's Sheldonian Theatre.

Newton, Isaac

Born	December 25, 1642 (Jan. 4, 1643, New Style)
Birthplace	Woolsthorpe, Lincolnshire, England
Died	March 20 (Mar. 31), 1727
Importance	Co-developer of calculus

Over and above his work on optics and calculus, Newton's laws of motion and gravity laid a foundation stone for modern physics—they were enough to map a route to the Moon 300 years later. With a childhood marked by the loss of his father and rejection by his mother, Newton the man was secretive, selfish, and vindictive. The fabled apple story is reputed to have happened while at the family home in Lincolnshire in retreat from the plague that was sweeping through cities. Newton guarded discoveries so jealously that it was often decades before he published them.

Pascal, Blaise

Born	June 19, 1623
Birthplace	Clermont-Ferrand, France
Died	August 19, 1662
Importance	Built early mechanical calculator

Blaise Pascal's genius was one that burned brightly from an early age but was seemingly self-extinguished in middle age. Pascal's calculators were begun while still a teenager, and his development of binomial theory, using his triangle of numbers, was fully complete by the age of 30. He had already made discoveries about vacuums that would eventually lead to the theory of relativity. However, in 1654 an intense religious vision resulted in Pascal ending his scientific career and devoting the rest of his life to theology.

Peano, Giuseppe

Born	August 27, 1858
Birthplace	Cuneo, Kingdom of Sardinia (now in Italy)
Died	April 20, 1932
Importance	Definition of axioms of math

As well as his work in the philosophy of math, this Italian mathematician is remembered for introducing the notations and symbols used in modern set theory. It was this field of math that Peano used to set out his axioms. After having rewritten Euclid's foundation of math, Peano set out to

supersede the *Elements* itself with Formulario Project, a compendium of formulae and theorems to date, all conforming to the same convention of notation. Peano then helped develop a whole new universal language based on Latin to spread mathematical concepts. It did not catch on.

Poincaré, Henri

Born	April 29, 1854
Birthplace	Nancy, France
Died	July 17, 1912
Importance	Leading figure in topology

Few figures are as worthy as Poincaré of the term polymath. He is most associated with the field of topology, which he helped to found (leaving his conjecture unsolved for most of the twentieth century), but he also worked on special relativity, quantum physics, gravity, what became chaos theory,  and electromagnetism. Poincaré, known as Jules to his family, was the brightest star in a glittering academic family. He spent a large part of his career as an inspector of mines, with math being a sideline (although a highly productive one).

Poisson, Siméon

Born	June 21, 1781
Birthplace	Pithiviers, France
Died	April 25, 1840
Importance	Leading figure in probability

While the statistical distribution that takes his name owes more of its success to the work of others, Poisson worked in a wide range of subjects. He was a child of the French Revolution, entering further education in 1798 after the worst of the political ructions had died down.

He devoted himself to science while post-revolutionary regimes rose and fell. He was a hard-working teacher and published more than 300 papers, many of them concerned with the application of math to problems in physics, such as magnetism and light. He was elevated to the position of baron but rarely used the title.

Pythagoras

Born	c.570 BCE
Birthplace	Samos, Ionia, Greece
Died	c.500–490 BC
Importance	Pythagoras theorem and math of music

There is no direct evidence of the life of Pythagoras. All we know of him comes from the accounts of others, in no small part from the writings of Plato. The facts, devotions, and philosophy are impossible to separate. Some scholars suggest that Pythagoras is the personification of a set of ideas rather than a person at all. Tradition dictates that Pythagoras the man was born on the island of Samos. He traveled widely—perhaps as far as India— absorbing the math of Babylon, Egypt, and beyond, before settling in Croton in southern Italy, where he formed his Pythagorean school.

Ramanujan, Srinivasa

Born	December 22, 1887
Birthplace	Erode, India
Died	April 26, 1920
Importance	Developments in number theory

The story of this Indian is one of raw genius. He had almost no formal schooling and he educated himself with the help of a couple of math students who lodged with his family as well as some borrowed books. By the age of 13 he began to develop his own theorems, discovering the work of the greats in his own way. He sent his workings to the world's mathematicians and was offered a place at Cambridge at the age of 27, where he worked on prime numbers. Although he survived smallpox as an infant, tuberculosis killed him by the age of 32.

Russell, Bertrand

Born	May 18, 1872
Birthplace	Trelleck, Monmouthshire, Wales
Died	February 2, 1970
Importance	Leading philosopher of mathematics

Born into a wealthy and influential British family that was active in the political elite from before the days of Henry VIII, Russell inherited an earldom. Despite his privileged upbringing, the young Bertrand was a lonely youth and considered suicide. However, he found his calling in math and philosophy, and had become a figure of world stature by his 30s. However, this was not the peak of his career. Russell was a staunch pacifist using his position to defend conscientious objectors in both world wars. He was also a leading anti-nuclear campaigner in the 1950s and '60s.

Riemann, Bernhard

Born	September 17, 1826
Birthplace	Breselenz, Hanover, Germany
Died	July 20, 1866
Importance	Founded elliptical geometry and zeta function

This German mathematician had a troubled childhood in a large, impoverished family. He lost his mother while still a child, and this left him painfully shy and phobic of public speaking. Allowances were made for this problem

during his academic career, but he was nevertheless required to give lectures in his job at the University of Göttingen. And many people were interested in what he had to say; his elliptical geometry inspired Einstein and his zeta function is the closest we have come to finding a pattern in prime numbers.

Stevin, Simon

Born	1548
Birthplace	Bruges, Flanders (now Belgium)
Died	1620
Importance	Co-founder of decimal system

The word mathematics is thought universal, barely changed from language to language. However, in Dutch the word is translated as *wiskunde*, meaning "the art of what is certain." This anomaly is thanks to Simon Stevin, a Flemish engineer and scientist who set out the terminology used. Another example is the Dutch word *middlellijn*, which means diameter. Stevin's other contributions include upgrading the water pumps and flood drains for the Netherlands and the invention of land yachts—wind-powered vehicles capable of high speeds.

Turing, Alan

Born	June 23, 1912
Birthplace	London, England
Died	June 7, 1954
Importance	Founding figure in digital computing

It is likely Alan Turing suffered from Asperger Syndrome, which makes it hard for people to empathize with others. Ironically he invented the Turing test for artificial intelligence—if a computer could fool a human into thinking it is also human, it passes the test. (No computer ever has.) Turing was a valued member of the British government's science community. When arrested for a homosexual act—a crime in the early 1950s—his security clearance was revoked. Without the high-level work to occupy him, Turing committed suicide by eating a poisoned apple.

von Neumann, John

Born	December 28, 1903
Birthplace	Budapest, Hungary
Died	February 8, 1957
Importance	Developer of game theory and computing

It was obvious from an early age that Janos Neumann was a clever child. He could speak ancient Greek at the age of six and divide eight-digit numbers into each other. He proceeded to be the youngest, brightest, and best at every institution he passed through from Budapest, Zurich, and Berlin, before arriving in Princeton in the 1930s, renaming himself John and working alongside Einstein and Gödel. von Neumann's game theory was a crucial weapon in the Cold War and the mathematician became involved in developing defensive strategies, such as ICBM missiles.

Viète, Francois

Born	1540
Birthplace	Fontenay-le-Comte, France
Died	December 13, 1603
Importance	Introduced symbols to algebra

Cynics who struggle with the xs, ys, and other mathematical language that Viète introduced might seek solace in the fact that he was a lawyer and cryptographer, and thus an expert in making things as complicated as possible and in hiding true meanings. Viète's code-breaking skills were given royal approval when he deciphered messages from the Spanish authorities, whose king was planning to depose France's Henry IV. With their plans revealed, the Spanish were so sure of their codes that they complained to the Pope that the French must have used black magic.

Wiles, Andrew

Born	April 11, 1953
Birthplace	Cambridge, England
Died	-
Importance	Proved Fermat's last theorem

Andrew Wiles had dreamed of solving Fermat's Last Theorem since the age of 10 after finding the puzzle in a library book while dawdling home from school. Born, raised, and educated in Cambridge (plus a few years at Oxford), Wiles made his career abroad in Paris and at Princeton. However, he chose to return to Cambridge to present his proof on Fermat's Last Theorem. Wiles received many awards for this momentous breakthrough, including being knighted by the British Queen, however he was just too old to win the greatest math prize of all, the Fields Medal (reserved for the under 40s).

BIBLIOGRAPHY AND OTHER RESOURCES

Books

Abbott, Edwin A. *Flatland*. New York: Dover Publications, 1992 [1884].

Bellos, Alex. *Alex's Adventures in Numberland*. London: Bloomsbury Publishing, 2011.

Biddiss, Mark. *Dr Mark's Magical Maths*. London: Hands On Publishing, 2004.

Boyer, Carl B. *A History of Mathematics*. Hoboken: John Wiley and Sons, 1991.

Clegg, Brian. *Infinity: The Quest to Think the Unthinkable*. London: Robinson, 2005.

Cooke, R. *The History of Mathematics*. New York: John Wiley and Sons, 1997.

Courant, Richard, Herbert Robbins and Ian Stewart. *What is Mathematics?* New York: Oxford University Press, USA, 1996.

Crilly, Tony. *50 Mathematical Ideas You Really Need To Know*. London: Quercus, 2007.

Ewald, William B., ed. *From Immanuel Kant to David Hilbert: A Source Book in the Foundations of Mathematics*. New York: Oxford University Press, USA, 1996.

Gardner, Martin. *The Colossal Book of Mathematics*. New York: W. W. Norton, 2001.

Hoffman, Paul. *The Man Who Loved Only Numbers*. New York: Hyperion, 1998.

Hogben, Lancelot. *Mathematics for the Million: How to Master the Magic of Numbers*. New York: W.W. Norton, 1983.

Linton, Christopher M. *From Eudoxus to Einstein–A History of Mathematical Astronomy*. Cambridge: Cambridge University Press, 2004.

MacArdle, Meredith (ed.). *Scientists: Extraordinary People who Changed the World*. London: Basement Press, 2008.

Maor, Eli. *"e," the Story of a Number*. Princeton: Princeton University Press, 2009 Russell, Bertrand. *Introduction to Mathematical Philosophy*. London: Routledge, 1993 [1919].

Scerri, Eric R. *The Periodic Table: A Very Short Introduction*. Oxford: Oxford University Press, 2011.

Singh, Simon. *Fermat's Last Theorem*. London: Fourth Estate, 1998.

––*The Codebook*. New York: Doubleday, 1999.

Stewart, Ian. *Does God Play Dice?* London: Penguin, 1989.

––*The Magical Maze*. New York: John Wiley & Sons, 1999.

Struik, Dirk J. *A Concise History of Mathematics*. New York: Dover Publications, 1987.

Mathematical Societies

African Mathematical Union www.math.buffalo.edu/mad/AMU

American Mathematical Society www.ams.org

Australian Mathematical Society www.austms.org.au

Austrian Mathematical Society www.oemg.ac.at

Brazilian Society of Mathematics www.sbm.org.br

Canadian Applied and Industrial Mathematics Society www.caims.ca

Canadian Mathematical Society www.ms.math.ca

Chinese Mathematical Society www.cms.org.cn

Clay Mathematics Institute www.claymath.org

Danish Mathematical Society www.dmf.mathematics.dk

Dutch Mathematical Society (Wiskundig Genootschap) www.wiskgenoot.nl

German Mathematical Society www.dmv.mathematik.de

Indian Mathematical Society www.indianmathsociety.org.in

Institute of Mathematics and its Applications (U.K.) www.ima.org.uk

International Mathematical Union www.mathunion.org

Italian Mathematical Union www.umi.dm.unibo.it

János Bolyai Mathematical Society (Hungary) www.bolyai.hu

London Mathematical Society www.lms.ac.uk

Mathematical Association of America www.maa.org

Mathematical Society of France smf.emath.fr

Mathematical Society of Japan www.mathsoc.jp

Polish Mathematical Society www.ptm.org.pl

Society of Mathematicians, Physicists and Astronomers of Slovenia www.dmfa.si

Spanish Royal Mathematical Society www.rsme.es

St. Petersburg Mathematical Society (Russia) www.mathsoc.spb.ru

Swedish Mathematical Society www.swe-math-soc.se

Swiss Mathematical Society www.math.ch

Union of Czech Mathematicians and Physicists www.cms.jcmf.cz

Union of Slovak Mathematicians and Physicists www.fmph.uniba.sk

General Museums

Arithmeum, Bonn, Germany. www.arithmeum.uni-bonn.de

China Science and Technology Museum, Beijing, China. www.cstm.org.cn

Computer History Museum, Mountain View, California, U.S.A. www.computerhistory.org

Copernicus Science Centre, Warsaw, Poland. www.kopernik.org.pl/en/

Deutsches Museum, Munich, Germany. www.deutsches-museum.de

DuPage Children's Museum, Naperville, Illinois, U.S.A. www.dupagechildrensmuseum.org

Erlebnisland Mathematik (Math Adventure Land), Dresden, Germany. www.math.tu-dresden.de

Exploratorium, San Francisco, U.S.A. www.exploratorium.edu

Federal Reserve Bank of Philadelphia, Money in Motion permanent exhibition, Philadelphia, U.S.A. www.phil.frb.org

Franklin Institute Science Museum, Philadelphia, U.S.A. www2.fi.edu

Goudreau Museum of Mathematics in Art and Science, New York, U.S.A. www.mathmuseum.org

iQ Park. Liberec, Prague, Czech Republic. www.iqpark.cz

Massachusetts Institute of Technology Museum, Cambridge, Massachusetts, U.S.A. www.web.mit.edu/museum

Mathematics Palace, Navet Science Center, Borås, Sweden. www.navet.com

Mathematikum, Giessen, Germany. www.mm-gi.de

MatheMuseumTirol, Innsbruck, Austria. www.mathemuseum.org

Museo Galileo, Institute and Museum of the History of Science, Florence, Italy. www.museogalileo.it

Museu de Matemàtiques de Catalunya, Barcelona, Spain. www.mmaca.cat

Museum of the History of Science, Oxford, U.K. www.mhs.ox.ac.uk

Museum of Mathematics, New York, U.S.A. (opening December 2012) www.momath.org

Museum of Natural Science and Scientific Instruments of the University of Modena, Modena, Italy. www.museo.unimo.it

Museum of Science, Boston, U.S.A. www.mos.org

MuseumsQuartier, Vienna, Austria. www.mqw.at

National Museum of Nature and Science, Tokyo, Japan. www.kahaku.go.jp/english

Norwegian Museum of Science and Technology, Oslo, Norway. www.tekniskmuseum.no

Observatory Museum, Stockholm, Sweden. www.observatoriet.kva.se/engelska

Ontario Science Centre, Toronto, Canada. www.ontariosciencecentre.ca

Palace of Discovery, Paris, France. www.palais-decouverte.fr

Pavilion of Knowledge, Lisbon, Portugal. www.pavconhecimento.pt/home

Powerhouse Museum, Sydney, Australia. www.powerhousemuseum.com

Science Museum, London, U.K. www.sciencemuseum.org.uk

Shanghai Science and Technology Museum, Shanghai, China. www.sstm.org.cn

Smithsonian Institution, Washington D.C., U.S.A. www.si.edu

Techniquest, Cardiff, U.K. www.techniquest.org

Archives and Individual Exhibits

Charles Babbage Papers and replica Difference Engine, Science Museum, London, U.K. www.sciencemuseum.org.uk

George Boole Archive, Royal Society, London, U.K. www.royalsociety.org. Also papers in the Boole Library, National University of Ireland, Cork Ireland; Trinity College, Dublin, Ireland; Cambridge University Library, Cambridge, U.K.

René Descartes Papers, Institut de France, Paris, France www.institut-de-france.fr

Albert Einstein Papers, Hebrew University of Jerusalem, Israel www.huji.ac.il

Enigma machines: Australian War Memorial, Canberra, Australia; Bletchley Park, Milton Keynes, England, U.K., Computer History Museum, Mountain View, California, U.S.A; Defence Signals Directorate, Canberra, Australia; Deutsches Museum, Munich, Germany; Museum of Science and Industry, Chicago, Illinois, U.S.A; National Cryptologic Museum, Fort Meade, Maryland, U.S.A.; National Signals Museum, Helsinki, Finland; Naval Museum of Alberta, Calgary, Canada; Polish Army Museum, Warsaw, Poland; Science Museum, London, U.K.; Swedish Army Museum, Stockholm, Sweden.

Kurt Gödel Papers, Princeton University, Princeton, U.S.A. www.princeton.edu

William Hamilton Papers, Trinity College, Dublin, Ireland www.tcd.ie. Also letters in Cambridge University Library, Cambridge University Library, Cambridge, U.K.; Royal Society, London, U.K.; British Library, London, U.K.

John Napier Papers, Lambeth Palace Library, London, U.K. www.lambethpalacelibrary.org

Isaac Newton Papers, Cambridge University Library, Cambridge, U.K. www. cudl.lib.cam.ac.uk

Bertrand Russell Papers McMaster University Library, Hamilton, Canada www.library.mcmaster.ca

Alan Turing Papers, King's College Archive Centre, University of Cambridge, Cambridge, U.K. www.kings.cam.ac.uk and National Archive for the History of Computing, Manchester University, Manchester, U.K. www.chstm.manchester.ac.uk

Websites

Khan Academy: www.khanacademy.org/#chemistry

MacTutor History of Mathematics archive: www-history.mcs.st-andrews.ac.uk

Nobel Foundation: www.nobelprize.org

App

Minds of Modern Mathematics. IBM, for iPad.

INDEX

© 2012 Shelter Harbor Press and Worth Press Ltd

All rights reserved. No part of this publication may be reproduced, stored in a retrieval system, or transmitted, in any form or by any means, electronic, mechanical, photocopying, recording, or otherwise, without prior written permission from the publisher.

Cataloging-in-Publication Data has been applied for and may be obtained from the Library of Congress.

ISBN 978-0-9853230-4-2

Series Concept and Direction: Jeanette Limondjian
Design: Bradbury and Williams
Copy Editor: Meredith MacArdle
Picture Research: Jennifer Veall
Cover Design: Jokooldesign

Publisher's Note: While every effort has been made to insure that the information herein is complete and accurate, the publishers and authors make no representations or warranties either expressed or implied of any kind with respect to this book to the reader. Neither the authors nor the publisher shall be liable or responsible for any damage, loss or expense of any kind arising out of information contained in this book. The thoughts or opinions expressed in this book represent the personal views of the authors and not necessarily those of the publisher. Further, the publisher takes no responsibility for third party websites or their content.

SHELTER HARBOR PRESS
603 West 115th Street Suite 163
New York, New York 10025

Printed and bound in China by Imago.

10 9 8 7 6 5 4 3 2

PICTURE CREDITS
BOOK

Alamy/The Art Archive 14; The National Trust Photolibrary 22 top; The Art Gallery Collection 22 bottom; Universal Images Group Limited 26 left; The Art Archive 28; Mary Evans Picture Library 29 top left; INTERFOTO 36; The Art Gallery Collection 37 bottom; INTERFOTO 39 top left; James Davies 40 top; INTERFOTO 42; Marc Tielemans 44 bottom; Universal Images Groups Limited 45; World History Archive 53 bottom; The Natural History Museum 55; Classic Image 58; Universal Images Group Limited 60 bottom; Steve Vidler 61 top; TravelCom 61 bottom; World History Archive 65 bottom; INTERFOTO 66; Universal Images Group Limited 67 left; Mary Evans Picture Library 69 left; Chris Howes/Wild Places Photography 69 right; INTERFOTO 70 top; Mary Evans Picture Library 76, 77; Pictorial Press Ltd. 80 top; Classic Image 82 top; INTERFOTO 84 top; Mary Evans Picture Library 88 right; INTERFOTO 90; Pictorial Press Ltd. 94 top; Mary Evans Picture Library 96; Pictorial Press Ltd. 98; INTERFOTO 100, 102 top; Tibor Bognar 104 bottom; Victor de Schwanberg 106; Peter Scholey 107 left; ITAR-TASS Photo Agency 166 bottom; Photos12 131 bottom left; INTERFOTO 131 bottom right; World History Archive 132 top right; INTERFOTO 132 bottom left; Adam Eastland Italy 133 bottom left; INTERFOTO 134 top left; North Wind Picture Archives 135 top left; Classic Image 136 top left; Mary Evans Picture Library 137 bottom left; Pictorial Press Ltd. 138 top right; INTERFOTO 138 bottom left; citypix 138 bottom right; Pictorial Press Ltd. 139 top left; The Art Gallery Collection 139 bottom left. **Bradbury and Williams** 10 top, 13 bottom left, 15 top, 16 bottom left, 17 top right, 23 bottom right, 24 left, 25 bottom right, 42 bottom left, 43 top right, 46 bottom right, 49, 50, 51 right, 52, 53 top right, 60 bottom, 62 bottom right, 67 bottom left, 71 bottom, 73, 79 bottom, 81, 83, 93, 108, 109, 110, 123, 124, 126, 127, 128, 129. **Clay Mathematics Institute** 116 top. **Corbis**/Werner Forman 12 bottom; Bob Sacha 26 right; Bettmann 44 top, 47; DK Limited 2-3, 48; Hulton-Deutsch Collection 134 top right; Baldwin H. Ward & Kathryn C. Ward 134 bottom right; 139 top right. **Getty Images**/Clive Streeter 4 bottom, 39 top right; DEA/R. Merlo 60 top; PNC; Time & Life Pictures 102 bottom. **Science Photo Library**/4 background; Royal Astronomical Society 5 background; Library of Congress, African and Middle Eastern Division 6; Middle Temple Library 7 left; George Bernard 12 top; 13 right; Bert Myers 16-17 centre; 18, 24 top; Sheila Terry 29 bottom left; Asian and Middle Eastern Division/New York Public Library 33; Royal Astronomical Society 34 left; American Institute of Physics 34 right; Middle Temple Library 37; CCI Archives 46; Middle Temple Library 53 top; Science Source 62; Royal Institution of Great Britain 74; 80 bottom; Middle Temple Library 82 bottom; 83, 85, 87 top; RIA Novosti 93 top; Royal Astronomical Society 94-95 bottom, 95 bottom; E.R. Degginger 97; National Physical Laboratory © Crown Copyright 99; Scott Camazine 105; Pasieka 111 right; Professor Peter Goddard 115; 117 top; Sheila Terry 133 top right; 134 bottom left; Emilio Segre Visual Archives/American Institute of Physics 135 bottom right; Royal Astronomical Society 136 top right; Royal Institution of Great Britain 137 top right; 138 top left; Professor Peter Goddard 139 bottom left. **Taken from Biblioteca Huelva/www.en.wikipedia.org/wiki/File:Aristoteles_ Logica_1570_ Biblioteca _Huelva. jpg 19. Taken from Jlrodri/www.en.wikipedia.org/wiki/ File:E8-with-thread.jpg 86. Taken from Pbroks13/http:// en.wikipedia.org/wiki/ File:Conic_sections_ with _ plane. svg 40-41. The Royal Belgian Institute of Natural Sciences, Brussels** 10 bottom. **Thinkstock**/iStockphoto 7 right, 11 bottom; Goodshoot 16 top; Hemera 21; Photos.com 25 top; John Foxx/Stockbyte 35; iStockphoto 38 top; Photos.com 38 bottom, 39 bottom, 41 top; iStockphoto 51 left; Photos.com 54; iStockphoto 56 right, 56-57 background; Photos.com 59, 64 top, 64 bottom, 68 top; Hemera 68 bottom; Photos.com 75 top; iStockphoto 78 bottom; Photos.com 79 top, 87 bottom; Hemera 88 left; iStockphoto 89 top; Hemera 89 bottom; Digital Vision 92; Hemera 95 top; iStockphoto 104 top; Comstock 105 top; iStockphoto 107 right; Stocktrek Images 111 left; Hemera 112; iStockphoto 114 top; Photos.com 114 bottom, 130 top right, 130 bottom right, 131 top right; iStockphoto 132 top left; Photos.com 132 bottom right, 133 top left; iStockphoto 133 bottom right, 135 top right; Photos.com 135 bottom left; Hemera 136 bottom left; iStockphoto 136 bottom right; Photos.com 137 bottom right. **United States Department of Agriculture** 23 left. **US Army Photo** 101. **Colin Woodman** 24 top, 55 bottom, 59 top left, 65 top right, 72 bottom, 74 top.

TIMELINES

Alamy/Classic Image; Danita Delimont; David South; Keystone Pictures USA; Images Group Limited; INTERFOTO; Mary Evans Picture Library; North Wind Picture Archives; Peter Jordan; Photo Art Collection (PAC); Pictorial Press Ltd.; Photos 12; Prisma Archivo; RIA Novosti; The Art Archive; UK Alan King; VIEW Pictures Ltd.; History Archive; ZUMA Wire Service. **Bradbury and Williams**. **Corbis**/Michael Ochs Archives. **Getty Images**. **Science Photo Library**/Dr. Jeremy Burgess; Mehau Kulyk; Sheila Terry. **Thinkstock**/Digital Vision; Fuse; Hemera; iStockphoto; Photos.com; Stocktrek Images; Top Photo Group. **Colin Woodman**.

Publisher's note: Every effort has been made to trace copyright holders and seek permission to use illustrative material. The publishers wish to apologize for any inadvertent errors or omissions and would be glad to rectify these in future editions.

CONTRIBUTORS by article

Richard Beatty 17, 37, 47, 58, 70, 76.

James Bow 10, 11, 18, 20, 63, 83, 87, 89, 91.

Dan Green 22, 42, 71, 75, 92, 99.

Mike Goldsmith 45, 46, 50, 51, 52, 54, 55, 59, 62, 67, 69, 72, 77, 80, 94, 95.

Tom Jackson 1, 2, 3, 4, 8, 9, 12, 13, 15, 16, 19, 21, 23, 32, 33, 35, 36, 41, 43, 53, 56, 61, 66, 73, 78, 81, 84, 85, 86, 88, 90, 93, 96, 97, 98, 100.

Robert Sneddon 5, 6, 7, 14, 24, 25, 26, 27, 28, 40, 44, 48, 49, 57, 60, 74, 79.

Susan Watt 29, 30, 31, 34 38, 39, 64, 65, 68, 82.

$$V_{ij}^{(1)} = \int U_i^{(0)+} \, V \, U_j^{(0)} \, d$$

$$\int$$

$$V_{12} \, \frac{1}{E^{(-1)} - \dot{H}_2} \, V_{12}^{+} \longrightarrow V_{12}$$

$$\boxed{\underline{E'}}$$

$$\langle \Phi_2^{(0)} | V$$

$$: E_2 + i \frac{\Gamma_2}{2}$$